By Jupiter

BY JUPITER

Odysseys to a Giant

Eric Burgess

Columbia University Press
New York
1982

Library of Congress Cataloging in Publication Data

Burgess, Eric.
By Jupiter

    Includes index.
    1. Jupiter (Planet) 2. Pioneer (Space
probe) 3. Project Voyager. I. Title.
QB661.B87   559.9'25 82-4139
ISBN 0-231-05176-X   AACR2

Columbia University Press
New York *and* Guildford, Surrey

Clothbound editions of Columbia University Press books are
Smyth-sewn and printed on permanent and durable acid-free paper.

To Chad and Brian

Acknowledgments

Particular thanks are due to the Public Information Offices of NASA's Ames Research Center and of the Jet Propulsion Laboratory of the California Institute of Technology, and to Richard O. Fimmel, Peter Waller, Frank Bristow, and Jurrie van der Woude. The author also expresses his gratitude for the many interviews with and comments from project personnel during these long missions to the outer giants of the Solar System. While many sources of information are gratefully acknowledged the views presented must, of course, be regarded as the author's own and not necessarily agreeing with official or other ideas on the subjects discussed.

Contents

By Jupiter

1.

Solar System Giant

WEALTHY Babylon, city of the hanging gardens—one of the seven wonders of the ancient world. Babylon, city ruled by the god Marduk, represented by the mighty planet Jupiter, the most splendid object of the night sky. While not quite as bright as Venus, Jupiter can dominate the heavens by riding high in the sky even at midnight. It is not bound close to sunset or dawn as is Venus. In the clear night skies of the Middle East, Jupiter is magnificently bright when in opposition to the Sun. Each year in Babylon at the great stepped ziggurat pyramid of Marduk, the temple became a ceremonial center for extravagant and colorful festivities. Babylonian astronomers devoted endless hours to observing Jupiter, measuring the apparent distances of the planet from bright stars and recording them in cuneiform texts. They tracked Jupiter's movements through the sky, and documented in clay tablets how each year the planetary wanderer made a great loop among the fixed stars as it traveled one-twelfth of its way around

the celestial sphere. Jupiter's annual motion (see box) was applied by the Babylonians to define the constellations of the zodiac. And when the planet was brilliantly opposite to the Sun in the star sphere, the Babylonians gave it the special name, Nibir. In the fifth tablet of the Tablets of Creation, the *Enuma Elish*, the Babylonians refer to this important role of Marduk: "He founded the station of Nibir to determine their bounds" (the boundaries for the zodiacal constellations).

In Roman and Greek mythology the bright planet was associated with the king of the gods. To the ancient Greeks it was Zeus and to the Romans, Jupiter. As with the other planets known to the ancients, we have retained the Roman name Jupiter as our modern name for the planet.

Ancient astronomers associated the planet Jupiter with their most powerful gods, and modern astron-

To Find Jupiter in the 1980s

From 1982 through 1990 Jupiter will progressively move through the zodiacal constellations from Libra to Cancer. For the first few years (1982–1986) the planet will appear as a bright but unwinking "star" low in the southern sky at midnight during the late spring and summer months. From 1987 to 1989 it will be higher in the southern sky during September through November. By the end of the decade it will be a winter planet riding high in the southern sky at midnight in December. For each hour before midnight the planet will be 15 degrees to the east of south, and for each hour after midnight, 15 degrees to the west. For each month before opposition it will be directly south 2 hours after midnight, and for each month after opposition it will be south 2 hours before midnight.

The other giant planet, Saturn, will move more sluggishly through the constellations from Virgo to Sagittarius in the same period. It is fainter than Jupiter and will appear in the midnight sky as a spring planet at the beginning of the period and a summer planet toward the end of the period.

The following table provides a guide to finding these giant planets:

Year	Month When South at Midnight	
	Jupiter	*Saturn*
1982	April	March
1983	May	April
1984	June	April
1985	July	May
1986	August	May
1987	October	June
1988	November	June
1989	December	June
1990	December	July

omers acknowledge Jupiter as the most important planet of our Solar System. It is certainly the largest, and the most massive, since it contains two-thirds of all the planetary matter in the Solar System.

Astronomers group planets of the Solar System into two types: small, dense, inner planets with solid surfaces (referred to as terrestrial-type planets), and large, mainly liquid, outer planets. Our Earth and Moon are in the first category, along with Mercury, Venus, and Mars. Jupiter is in the second category, as are Saturn, Uranus, and Neptune. These large outer planets have families of satellites, some of which are as big as the smaller inner planets. Pluto, the outermost known planet, is difficult to classify. We cannot observe it in detail because of its small size and great distance.

A zone between the orbits of Mars and Jupiter, dividing the small inner planets from the outer giants, contains a wide belt of asteroids or minor planets. The largest, Ceres, is about 635 miles in diameter. Others are very much smaller, and many are irregularly shaped, as though fragmented from larger bodies.

As spacecraft allowed close looks at the inner bodies of the Solar System and revealed them as such diverse worlds, scientists and space technologists became very interested in missions to the outer planets, particularly to giant Jupiter.

Jupiter, the Dominant Planet

Compared with Earth (figure 1-1), Jupiter is quite unusual. Only slightly denser than water, it contains 317.8 times as much mass as Earth. Truly dominating the Solar System with its enormous gravity, Jupiter strongly perturbs all the other planets in their orbits, and may have prevented material in the zone of the asteroids from forming into a single planet. Cometary orbits are distorted by Jupiter, and some comets with short orbital periods are controlled by the giant planet so that the aphelion (most distant point in

Figure 1-1: Compared with Earth, Jupiter is a giant planet. Here an Apollo picture of Earth is compared with a Pioneer image of Jupiter to the same scale. Note the relative sizes of the Great Red Spot and Earth and the size of the shadow of the satellite Io (the black spot on Jupiter).   (JPL/NASA)

cooling of the planet following its primordial gravitational collapse 4.5 billion years ago, soon after the formation of the Solar System.

The Satellite Worlds of Jupiter

As long ago as the thirteenth century learned men such as Roger Bacon wrote about instruments consisting of a combination of lenses which made distant objects appear near. But not until early in the seventeenth century did telescopes become common. Several Dutch opticians made and sold the instruments. The most well known was Hans Lippershey of Middleburgh, Holland, who in 1608 used a convex and a concave lens at opposite ends of a tube to construct the first practical telescope for military purposes. The news spread quickly and within a year two men in different European countries had heard about the new instrument: Simon Meyer (Marius) of Anspach, Germany, and Galileo Galilei of Padua, Italy, both interested in science and the heavens. Simon Marius obtained a Dutch telescope and Galileo built his own instrument. When each looked at Jupiter through his telescope he saw that the bright planet was not starlike but appeared to be a disk, and even more surprising, it possessed a system of satellites. That a planet other than Earth should have satellites ran counter to all accepted thought based on the Earth-centered philosophy of Aristotle. Critics of the new discoveries and their discoverers stated that the luminous objects flitting from side to side about Jupiter as reported by Marius and Galileo were quite obviously defects of the new instrument, not real objects.

Galileo is usually credited with discovering the four big Jovian satellites. He observed three of them on January 7, 1610, and the fourth a few days later. They are often referred to as the Galilean satellites. However, some historians say that it was Marius who saw the satellites first, on December 29, 1609. Unfortunately, Marius did not publish his observation until after Galileo. But it was Marius who named the satellites by the names we use today: Io, Europa, Ganymede, and Callisto—the mythological lovers of Jupiter. The satellites are also referred to by the Roman

orbit from the Sun) of each cometary orbit is located at about the distance of the orbit of Jupiter.

In spite of the fact that Jupiter is a huge planet it is not large enough to have become a star. Its weight was insufficient to raise internal temperatures high enough for nuclear reactions to start releasing energy by building hydrogen into helium. However, if Jupiter had been about 100 times its size, our Solar System might have had a second star and become a double star system. Then evolution of life on Earth might have followed a very different path. Earth would have had few nights; either the Sun or an equally brilliant but smaller Jovian sun would have been shining in the skies of our planet. Only for a few months each year would the two stars have appeared close together in the day sky so that there could be nighttime on our planet. Even today Jupiter emits several times more energy than it absorbs from sunlight, energy which must be derived internally and probably arises, at least partly, from continued

numerals I, II, III, and IV, respectively. Galileo wanted to call them the Medicean planets in honor of his sponsor the Duke of Medici.

The orbital motions of the satellites led to another important discovery. Ole Roemer was a Danish astronomer who in 1675, while at the Paris Observatory, explained puzzling discrepancies found in astronomers' observations of the eclipses and occultations of the Jovian satellites. An eclipse occurs when a satellite passes through the shadow cast by Jupiter. An occultation takes place when a satellite goes behind Jupiter as viewed from Earth. Astronomers had observed that eclipses and occultations of Jovian satellites occur increasingly later as Earth moves away from Jupiter. When the two planets are on opposite sides of the Sun the delay is 16 minutes and 40 seconds. Roemer explained that this delay results from the finite velocity of light. Light traveling across Earth's orbit, when Earth is farthest from Jupiter, takes 16 minutes and 40 seconds more to cover the additional distance than when Jupiter and Earth are at their closest, i.e. Jupiter in opposition. Roemer showed that the velocity of light is about 186,000 miles per second.

The Galilean satellites of Jupiter are planet-sized bodies (figure 1-2). Callisto and Ganymede are approximately the size of Mercury. Io and Europa are about the same size as our Moon. In a small telescope the four satellites appear as starlike objects in a nearly straight line on either side of the disk of the planet. This is because their orbits are viewed from Earth almost edgewise.

Io, the innermost of the big satellites, has a bright surface with a distinct orange color. It reflects five times as brightly as the surface of Earth's Moon, and is bright in infrared also. Io has nearly the same density as Mercury, so astronomers regard it as a rocky planet. There was no evidence of ice on its surface, but spectroscopes revealed a cloud of sodium vapor around the satellite.

Europa, the smallest of the Galilean satellites, is less dense than Io and was thought to be a rocky object.

Figure 1-2: Comparisons of the Galilean satellites with Mercury, the Moon, and Titan (the large satellite of Saturn).
(NASA)

Strangely, however, water was detected as ice and frost covering the whole surface of the satellite. The surface is, however, reddened and darkened away from the poles and on the trailing hemisphere. (All the satellites have one hemisphere turned toward Jupiter.)

Ganymede is the giant satellite of the Jovian system, sized between Mercury and Mars. Its low density indicates that it is mainly water with perhaps a small rocky core. The surface appears to be covered with frost, but there are some rocky areas.

The outermost Galilean satellite, Callisto, also has low density but astronomers could not find evidence of much frost or ice on its surface. One explanation was that the satellite had never melted and differentiated so that it consists of a well-mixed rubble of rock and ice out of which impacting meteorites had made a rocky regolith on the surface.

A fifth satellite of Jupiter, Amalthea, was not discovered until almost three centuries after the time of Galileo. Today, Jupiter is known to have at least sixteen satellites—the others are much smaller bodies than the Galilean satellites. The Jovian system thus resembles a small Solar System. But the outermost four satellites of Jupiter orbit oppositely to the others. The others revolve around Jupiter as all the planets go round the Sun—in one direction, counterclockwise as seen from a point above the north pole of the Sun.

Jupiter in the Family of Planets

The ancient astronomers of Babylon, observing the motions of the planets against the background of stars, called them wandering stars. Our modern word "planet" is derived from a Greek word meaning "wanderer." Today we know that all the planets, including the Earth, travel around the Sun in orbits that are close to circular. Mercury and Venus, orbiting the Sun within the orbit of Earth, are referred to as inferior planets. Mars, Jupiter, Saturn, Uranus, Neptune, and Pluto, orbiting outside the orbit of Earth, are superior planets. As seen from Earth all planets appear to move eastward close to the ecliptic—the apparent yearly path of the Sun relative to the constellations which is the projection of the plane of the Earth's orbit, the ecliptic plane, on the celestial sphere as seen from Earth. The ecliptic passes through the twelve constellations of the zodiac.

Jupiter takes 11.86 Earth years to orbit the Sun. As viewed from Earth, Jupiter follows a path through the fixed stars approximately along the ecliptic, progressively passing to the east through the zodiacal constellations as first recorded by the Babylonians.

When a superior planet is directly opposite the Sun in the sky it is in opposition—at that time Earth is between the Sun and the planet. At opposition the planet is also at its closest to Earth and accordingly appears at its brightest in Earth's sky. Jupiter reaches opposition approximately every 13 months.

Conjunction occurs when the planet is on the part of its orbit directly behind the Sun as seen from Earth. At the time of conjunction the planet viewed from Earth appears so close to the Sun as to be unobservable. It is then most distant from Earth.

The Earth moves faster in its orbit than any superior planet. As a consequence, around the date of opposition there is a period when a superior planet is being overtaken by the Earth. The result is that the superior planet appears to move backward—toward the west—relative to the stars. This is referred to as retrograde motion. Because the planes of planetary orbits differ slightly from that of Earth's orbit the observed path of the planet relative to the stars is usually a large flattened loop. The closer the planet, e.g. Mars, the larger is the loop it makes at opposition. A typical loop for Jupiter is shown in figure 1-3.

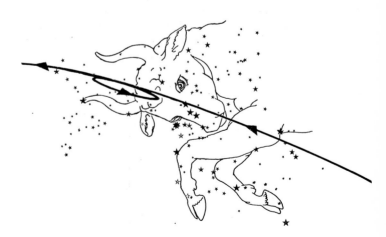

Figure 1-3: At each opposition (approximately every 13 months) Jupiter loops through the constellations of the Zodiac. This drawing illustrates a typical loop in the constellation of Taurus, the Bull.

An Earth-based View of Jupiter

Jupiter is the fastest rotating planet of the Solar System; it turns on its axis in 9 hours and 55.5 minutes. But all latitudes do not rotate at the same rate; equatorial regions, for example, rotate faster—in 9 hours. Jupiter's equatorial diameter is 11 times that of Earth. Because of its rapid rotation, the giant planet is quite flattened see figure 1-6); it measures 82,967 miles from pole to pole and 88,734 miles equatorially.

Jupiter's volume is 1,317 times that of Earth but its mass is just under 318 times Earth's mass. Its density is thus quite low compared with Earth's—1.314 gm/cc compared with Earth's 5.5 gm/cc. With such a low density Jupiter cannot be a solid body; it consists mainly of gas and liquid with possibly a core of molten rock. Jupiter's composition, predominently hydrogen and helium, is more like that of the Sun than that of Earth.

Seen from Earth through a telescope, Jupiter presents a magnificent sight—a striped, banded disc of turbulent clouds with all the stripes parallel to the planet's equator. Large, dusky gray regions cap each pole in amorphous hoods. Dark bands are called belts; lighter bands between the belts are called zones.

Jupiter's clouds are not brilliantly colored, but their colors are quite definite; they also change from year to year, and their patterns change even more frequently (figure 1-4). Zones vary from yellowish to white, while belts vary from gray to reddish brown. As well as changing color, the bands also fade and darken, widen or become narrow, and move up and down in latitude. But they have sufficient permanence to be given names (figure 1-5).

Many smaller features, such as arches, festoons, loops, plumes, spots, streaks, and wisps, add tantalizing details to the clouds of Jupiter as they change form rapidly within a few hours or a few days. Although difficult to resolve in great detail from Earth, these features in the past were generally accepted as

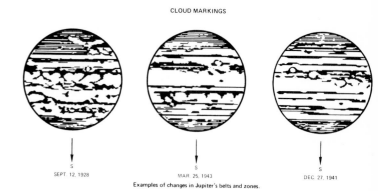

CLOUD MARKINGS

SEPT. 12, 1928   MAR. 25, 1943   DEC. 27, 1941

Examples of changes in Jupiter's belts and zones.

Figure 1-4: Jupiter belts and zones are regions of clouds that change quite considerably from time to time. These comparative drawings show telescopic views of the planet in 1928, 1941, and 1943.   (NASA)

being groups of clouds or turbulent regions between jet streams moving at different speeds.

By observing some of these details astronomers discovered that cloud features move around Jupiter at different rates. A great equatorial current, forming a 20-degree-wide girdle around the planet, travels 225 mph faster than cloud regions on either side.

A unique feature on Jupiter mystified generations of astronomers. It was first seen during the reign of Charles II, about the time when that king granted some English noblemen a large territory which extended south of Virginia to the Spanish possessions in Florida and later became the Carolinas. The dis-

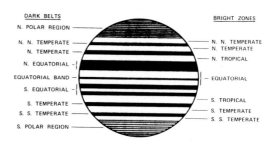

DARK BELTS   BRIGHT ZONES

N. POLAR REGION

N. N. TEMPERATE   N. N. TEMPERATE
N. TEMPERATE   N. TEMPERATE

N. TROPICAL

N. EQUATORIAL

EQUATORIAL BAND   EQUATORIAL

S. EQUATORIAL

S. TROPICAL

S. TEMPERATE   S. TEMPERATE
S. S. TEMPERATE   S. S. TEMPERATE

S. POLAR REGION

Figure 1-5: The regions of belts and zones are sufficiently permanent on the average for them to be given the names shown in this NASA diagram.

covery of the strange feature on Jupiter was made in England by Robert Hooke, who had just been appointed to the newly formed Royal Society and was working hard to perfect a chronometer to determine longitude at sea. He was also interested in astronomy, and when observing Jupiter he saw in the planet's southern hemisphere a long, oval, red feature. Known as the Great Red Spot (figure 1-6) it has remained a permanent feature of Jupiter since that time, although it sometimes fades and several times it has become virtually invisible. In the 1860s, for example, it was a faint oval ring, yet by the 1870s it was the most conspicuous feature on Jupiter and measured some 30,000 miles long by 7,000 miles wide, its color a deep brick red. In 1883 it was almost invisible again, then it revived, but by 1908 it was again almost invisible. Today it is quite conspicuous and about 15,000 miles long. Before spacecraft flew by Jupiter there were many speculations about the nature of the Great Red Spot, ranging from its being a high mountain peak to an island floating in the clouds. Jupiter also has small and less persistent red spots and relatively short-lived white spots. It re-

Figure 1-6: The polar flattening of Jupiter and the Great Red Spot that mystified generations of astronomers are shown clearly in this Earth-based telescopic photograph of the giant planet.   (NASA/Ames)

mained for spacecraft to solve the mystery of the Jovian spots, as described later.

Except for the Sun, Jupiter is by far the most intense source of radio signals in the Solar System. Before any spacecraft went to Jupiter, radio astronomers had observed radiation from the planet at three wavelength bands; thermal, decimetric, and decametric. Thermal radio waves are produced in the atmosphere of the planet. Decimetric radio waves originate from electrons spiraling along magnetic field lines outside the Jovian atmosphere. Decametric radio waves are produced by electrical discharges.

Decametric emissions were detected in 1954. They had been observed in 1950 but had not been recognized as coming from Jupiter. Thermal radio waves from Jupiter were first measured in 1956. Jovian decimetric radio emission was discovered in 1958, about the time that James Van Allen discovered Earth's radiation belts. Scientists concluded that the decimetric radio waves must originate from Jovian radiation belts similar to those of Earth in which charged particles are trapped and move backward and forward along magnetic field lines. And if this is so, they reasoned, Jupiter must possess a magnetic field which is many times stronger than that of Earth.

One strange aspect of the radiation from Jupiter is that it is periodic; there is approximately a 10-hour decimetric period associated with the rotation of Jupiter and a decametric period associated with the orbital period of the satellite Io, the innermost of the big satellites. The reason for such modulations and other questions about these radio emissions from Jupiter puzzled radio astronomers during the 1960s and stimulated interest in a space mission to the giant planet.

Interior of Jupiter

For many years scientists accepted that the interior of Jupiter must be quite different from that of Earth and the other terrestrial-type planets (figure 1-7). The

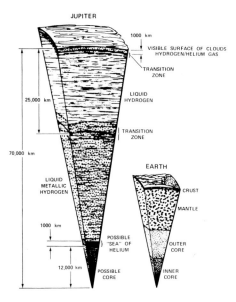

JUPITER

1000 km

VISIBLE SURFACE OF CLOUDS
HYDROGEN/HELIUM GAS

TRANSITION
ZONE

LIQUID
HYDROGEN

25,000 km

TRANSITION
ZONE

70,000 km

EARTH

LIQUID
METALLIC
HYDROGEN

CRUST

MANTLE

1000 km

POSSIBLE
"SEA" OF
HELIUM

OUTER
CORE

12,000 km

POSSIBLE
CORE

INNER
CORE

Figure 1-7: The interiors of Earth and Jupiter as now understood are shown in this drawing. While Earth consists principally of silicate rocks with a core of metals, Jupiter is mainly a hydrogen planet.   (NASA/Ames)

dynamic changes we observe through telescopes are cloud patterns in the planet's atmosphere. This outermost shell of Jupiter consists mainly of the cosmically abundant gas hydrogen, with some helium and traces of other gases including methane, ammonia, and water vapor. The cloud tops of Jupiter are at a temperature of about −190°F but temperature increases with depth. At about 600 miles below the visible cloud tops, it is 3600°F, but at that depth pressure is so great that hydrogen can no longer exist as a gas, it becomes liquefied. A thick shell of liquid hydrogen below the atmosphere extends to great depths, but there is another level of significant change. It is about 15,000 miles below the cloud tops, where the temperature is about 20,000°F and the pressure is three million Earth atmospheres. Liquid hydrogen changes into a special form called liquid metallic hydrogen, which behaves as a metal and readily conducts heat and electricity.

Jupiter's magnetic field is probably produced by seething convective motions within this shell of metal-

lic hydrogen, caused by heat—some internally generated and some residual from the formation of the planet—moving outward toward the surface of the planet.

Near the center of Jupiter, slushy masses of water, ammonia, and methane might surround a core of molten rock and metallic materials, perhaps as much as 20 times the mass of the whole Earth. Some theoretical models of Jupiter's interior suggest that helium separates from hydrogen to form a 620–mile thick shell surrounding the core. The core itself is believed to be very hot because of the tremendous pressure there. Its temperature may be as high as 54,000°F.

At some levels within the atmosphere of Jupiter temperatures are the same as those experienced on Earth. At such levels ammonia crystals could change to liquid ammonia droplets and water could also condense to rain from the clouds, sometimes frozen into snows of water and ammonia. But the drops and snowflakes cannot fall to a surface as they do on Earth. Instead, when they reach warmer lower regions in the atmosphere, they evaporate and the vapor rises back into the clouds.

This circulation pattern is roughly analogous to those that build up violent thunderstorms and tornadoes in Earth's atmosphere. More violent by far than any thunderstorms on Earth, the Jovian storms could give rise to electrical discharges that would make Earth's lightning flashes seem mere sparks. Such lightning flashes cannot be seen from Earth, but they might be detected from a spacecraft flying by Jupiter. Jet circulations in the cloud bands appear analogous to Earth's major atmospheric patterns, such as the trade winds, tropical convergences, and jet streams.

A Question of Living Things

One would think that Jupiter would be an inhospitable planet to life of any kind, but this may not be true. Since there are probabaly liquid water droplets in Jupiter's atmosphere of hydrogen, methane, and ammonia, the atmosphere of the giant planet might provide the same kind of primordial "soup" in which, according to the theories of some scientists, life originated on Earth.

Probably about 3.5 billion years ago carbon-based molecules became organized on Earth into living systems able to reproduce themselves. The current biological theory is that from then on, through a sequence of slight changes to subsequent copies, biological evolution produced all the living terrestrial creatures, including ourselves.

In 1953, Stanley Miller, a student of Nobel Laureate Harold Urey, simulated bolts of lightning in the kind of atmosphere retained by Jupiter by discharging electricity through a mixture of hydrogen, methane, ammonia, and water vapor. Some of the gas molecules combined into more complex molecules of the type believed to be the building blocks of living systems. This appeared to be a partial confirmation of those theories that organic life could have evolved in the atmosphere of Jupiter.

Some biologists were intrigued by the possibility of life on Jupiter. The temperature in vast regions of Jupiter's atmosphere is right, the gas mixture is suitable, and electrical discharges take place. The big question was whether or not a life form would have time to survive and evolve before being swept into regions of the atmosphere that are hot or cold enough to destroy it.

While none of the early missions to Jupiter were instrumented to search for life in its atmosphere, there are plans to send other missions that will include probes to penetrate the atmosphere. While there are no plans to include life-detection experiments in such probes, the intriguing possibility remains that the probes will confirm that conditions are, indeed, favorable to life in Jupiter's atmosphere even though detection of living things is extremely difficult. The problem is that we can only base a search on our form of life because it is the only form of life of which we are aware.

Why Jupiter?

Why a mission to Jupiter? The question of how our Solar System formed and gave rise to the Earth and to ourselves has always intrigued thinkers. Most civilizations of note have tried to find answers through mythology, religion, philosophy. Now we have the opportunity to add meticulous scientific observation at a level of sophistication never before approached in the entire history of our species. Today we are still far from having satisfactory answers, but we have added considerably to our knowledge through exploration of the nearby planets of the Solar System and observations of our Earth from space. Big questions remain as to how the various planets evolved to their present unique and different states and how life originated and flourished on Earth, the planet that differs so much from all others.

We cannot find complete answers here on Earth because we are looking at one clear snapshot in an album of pictures where all the earlier pictures are blurred, distorted, or missing. This single picture does not provide enough information for us to be really sure about Earth's past. And it provides little confidence in projecting Earth's future. But we believe that other planets may pass through evolutionary phases at different rates, so that they will provide other pictures which we can relate to Earth's past and possibly its future too.

All the planets are much too far away to be observed in meaningful detail by use of telescopes located on Earth. Since planetary probes have observed other planets at close range, virtually all generally accepted ideas about the planets have had to be changed. Without exception every exploration of a planet by a spacecraft produced surprises of such a radical nature that all existing astronomical textbooks became completely erroneous in their information about the planets. In fact, during the first few years of the space age we learned more about the planets than in all the previous centuries of observations from Earth.

This knowledge is vital to our understanding of the past and the future of Earth. Without such knowledge we can never be sure how to protect Earth's environment from irreversible damage.

In many respects giant planets such as Jupiter provide us with a model of what might be occurring generally in the universe. Many processes taking place within Jupiter may be analogous to those taking place within stars prior to their nuclear ignition. Additionally, the enormous magnetosphere of a planet such as Jupiter, with its strong magnetic field, provides new insight into the behavior of plasmas which cannot be duplicated in terrestrial laboratories.

The outer Solar System was relatively unknown to Man before the 1970s, when a new breed of spacecraft made journeys to Jupiter and on to Saturn. Yet the big planets of the outer Solar System are of great importance to our developing an improved understanding of the system's origin. The satellite system of Jupiter, for example, represents a small version of a solar system. The density of the Galilean satellites decreases with increasing distance from the central body, suggesting that their formation may have paralleled the formation of the Solar System.

To Explore the Outer Solar System

Jupiter is a veritable gateway to the outer Solar System. Since the planets of the outer Solar System are so distant, and government financial support for long-term scientific projects is difficult to obtain and sustain in the economic climate of the last quarter of the twentieth century, these planets can only be explored by spacecraft that travel fast enough to reach them in reasonable time. We have no launch vehicles available to accelerate to these high speeds spacecraft which are capable of carrying suitable payloads of scientific instruments. However, if a spacecraft is navigated carefully to pass close to Jupiter, the gravitational field and orbital motion of the giant planet can be used in a slingshot technique to hurtle the spacecraft relatively quickly to the outer

planets. Nevertheless, the required high velocities demand that even when slingshot techniques and the best available launch vehicles are used, an outer planets spacecraft must have a lightweight design for its structure and also for its components and scientific instruments.

Although Jupiter offers a means to explore the outer Solar System, using the planet in such a way was not free of problems. From analysis of the radio waves emitted by Jupiter, scientists knew that charged particles must be trapped by Jupiter's strong magnetic field to produce radiation belts around the planet, possibly extending farther from Jupiter than the distance between Earth and Moon. Without exploring these radiation belts, scientists could not be sure that the belts would not damage a spacecraft passing close enough to Jupiter to use that planet as a gravity slingshot to the outer worlds. This was a very important question that needed an answer, for if the radiation belts proved to be a serious hazard, the exploration of the outer Solar System might have to wait several decades until practical propulsion systems could be developed with higher specific impulses than those of chemical rockets.

Basically the problem was how many protons might be present in the radiation belts of Jupiter, because protons are the particles that would damage electronic components of a spacecraft. The radio waves from Jupiter carried information that told scientists approximately how many electrons were trapped in the belts, but there was no way of finding out by observations of Jupiter from Earth how many high-energy protons were in the belts. To obtain such information a spacecraft had to plunge through the radiation belts of Jupiter and measure the particle flux and composition there. This was a primary scientific objective of the first Jupiter mission—Pioneer Jupiter. It would pave the way for more sophisticated spacecraft such as Voyager and Galileo, which would be launched in later years if the pioneering mission were successful.

But a pioneering mission to Jupiter posed many technical challenges. It would extend Man's exploration of the Solar System to an unprecedented scale. While interplanetary journeys to Venus, Mercury, and Mars covered distances from the Sun of only tens of millions of miles, to explore the outer Solar System required traveling hundreds of millions of miles. The first mission to Jupiter would carry a spacecraft outward from Earth some half a billion miles, with possibilities of exploring interplanetary space to Saturn and then beyond the orbits of Uranus and Neptune. Though the spacecraft would not approach close to the latter planets, it could perhaps maintain communication with Earth from a distance of several billion miles.

Such vast distances present enormous problems to communications engineers. First, radio signals received over these distances are whisper faint, almost drowned by radio interference from other sources and by electronic noise in the components of the receiving apparatus. Second, radio waves travel at a finite velocity, the speed of light—186,000 miles/sec. So there is a time delay for information transmitted from the spacecraft to arrive at Earth, and an equal time delay for radio commands from Earth to arrive at the spacecraft. Such delays make it mandatory for controllers on Earth to operate the spacecraft and all its systems 90 minutes in advance of commanded events at the distance of Jupiter. Every contingency had to be planned for; there was no margin for error, no opportunity to change commands or operations in less time than 90 minutes.

Another problem arose from the immense distances involved; that of providing electrical power for the spacecraft. When operating in the inner Solar System spacecraft obtain their power by converting solar radiation into electricity via solar cells mounted on arrays large enough to provide the necessary collecting area. At Earth's orbit sunlight delivers about two calories per square centimeter of collector. Inefficiencies of conversion reduce the available electrical power considerably, but it is still sufficient for spacecraft in the inner Solar System. At the distance of Jupiter, however, sunlight carries only one twenty-seventh of the energy available at Earth's distance from the Sun. Solar cells to tap this reduced energy

source would cover too much area to be practical. So a spacecraft bound for the outer Solar System must use a nuclear energy source to generate electricity.

Another problem was the time involved in missions to the outer Solar System. Each spacecraft has to operate in space for many years before its mission can be completed. New levels of high reliability had to be reached.

There was an astronomical problem, too, in the environment of the Solar System itself. Between Mars and Jupiter is the asteroid belt. Some theories suggested the asteroid belt was a 175-million-mile-wide zone of abrasive dust the particles of which would impact and seriously damage any spacecraft trying to travel to the outer Solar System.

Figure 1-8: Pioneer was the first mission to the outer planets. This photograph shows Charles Hall (center), project manager for Pioneer, John Wolfe (right) project scientist, and Peter Waller, public affairs officer for Pioneer, examining a model of the spacecraft at NASA Ames Research Center.

(NASA/Ames)

These were some of the major obstacles. But the opportunity to explore the outer Solar System beyond the orbit of Mars beckoned strongly, challenging our ingenuity. Some forward-thinking scientists under the chairmanship of James Van Allen, discoverer of Earth's radiation belts, and a long-time advocate of rocket exploration of the environment beyond Earth's atmosphere, accepted the challenge. They recommended to the National Aeronautics and Space Administration that the nation make an exploratory mission to Jupiter. NASA later authorized two spacecraft, Pioneers F and G (figure 1-8), for the first journey to the outer Solar System. Their mission was truly a pioneering odyssey into vast regions of previously unexplored space and to the giant planets. Early in 1970 the story began to unfold of an epoch-making journey to the planet Jupiter and beyond, a mission to a spectacular object in the night skies of Earth that had occupied many generations of Babylonian astronomers and, in the age of the telescope, held the attention of modern man for centuries. The odyssey to Jupiter was a mission that promised to open a doorway to the outer regions of our Solar System.

2.

Pioneer Odysseys

MAY 1970: the conference auditorium in the Main Administration Building at NASA Ames Research Center, Mountain View, California, was filled to overflowing. Hundreds of engineers and scientists had crowded there to hear plans for two exciting missions to Jupiter. The conference was the first of a series in which scientists would be planning a strategy for exploration of the giant of the Solar System.

There was great scientific interest expressed in Man's first venture beyond the orbit of Mars. The stimulating goal was to assess hazards in deep space and to develop technology and deep-space operational experience for Grand Tour missions to the outer planets then being optimistically planned for the late 1970s. To explore the mysterious worlds of the outer regions of our Solar System seemed the logical and natural step following the outstanding successes of the Apollo manned expeditions to the Moon, and the unmanned exploration of the planets of the inner Solar System.

The Grand Tour derived from a report published by the Space Science Board of the National Academy of Sciences and the National Research Council, *The Outer Solar System: A Program for Exploration.* James A. Van Allen, who in the late 1940s spearheaded the exploration of Earth's upper atmosphere by launching rockets from balloons to gather data about the new frontier of space, was chairman of the study. His sights were now aimed at the outer Solar System. The 1969 study showed that several of the large outer planets could be visited on a single space flight in the late 1970s due to the unusual positions of the planets at that time. Such a mission would substantially reduce the time and cost involved in exploring the outer Solar System. Equally favorable planetary conditions would not occur again for nearly two centuries. It was a golden opportunity for the people of the United States to use their advanced technology and expertise acquired during the early years of the space program to complete the exploration of our Solar System. The study recommended missions that would first explore Jupiter, to be followed by two Grand Tour missions; an Earth-Jupiter-Saturn-Pluto mission in 1977, and an Earth-Jupiter-Uranus-Neptune mission in 1979.

But the story of the odysseys to Jupiter started much earlier, almost at the very beginning of our space program. In 1957 the Advanced Research Projects Agency authorized the launching of small unmanned spacecraft in an attempt to place a scientific payload near the Moon—our first attempt to reach another world in space. The small spacecraft were called Pioneers, and they were the first probes into interplanetary space (figure 2-1).

In May 1960, about a year after NASA's formation, an informal study began at NASA's Ames Research Center. This study, under the leadership of Charles F. Hall, was intended to demonstrate the Center's ability to manage space projects. The study recommended a small, cone-shaped spacecraft as a simple and long-lived spacecraft to explore the interplanetary medium within Earth's orbit. The then Director of the Center, Smith J. DeFrance, supported these new activities at the Center, which had previously been a laboratory of the National Advisory Committee for Aeronautics. In September 1960 he authorized a formal team and in the following years this team, together with interested scientists, tried to stimulate enthusiasm for the concept at NASA Headquarters.

The result was that Edgar M. Cortright, who was then Deputy Director of the Office of Space Science at NASA Headquarters, suggested that the team become involved in an Interplanetary Pioneer spacecraft. Space Technology Laboratories, which had been involved in the first Pioneer spacecraft for the Air Force, was chosen to complete a feasibility study for the spin-stabilized spacecraft concept they had evolved, and the company was later selected to build the spacecraft.

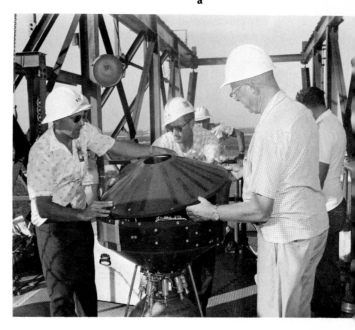

a

Figure 2-1: The first spacecraft in the Pioneer series was launched into space at the beginning of the space age in an attempt to send an instrumented package to the vicinity of the Moon.

(a)   Pioneer 1 is placed on the fourth stage of the Thor Able-1 launch vehicle.   (US Air Force)

(b)   Air Force lunar probe Thor Able-1 launches the first Pioneer into space at 4:42 a.m, October 11, 1958.

(US Air Force)

(c)   Comparison drawings of early Pioneers.   (NASA/Ames)

(d)   Pioneer 7 is prepared for launching atop a Thor-Delta into an orbit around the Sun, August 1966.   (NASA)

(e)   The Atlas/Centaur launch vehicle for the 1973 Pioneer mission to Jupiter, showing how the spacecraft was housed in a nose fairing (shroud) for launch.   (NASA/Ames)

Pioneers 6 through 9 were launched by Thor-Deltas between 1965 and 1968 to explore interplanetary space for several million miles inside and outside of Earth's orbit. They operated successfully for several years, gathering new information about the solar wind, the extent of solar cosmic rays, the structure of the Sun's plasma and magnetic fields, the physics of particles in space, and the nature of solar flares.

b

d

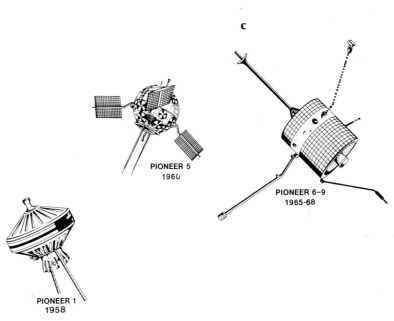

c

PIONEER 5
1960

PIONEER 6-9
1965-68

PIONEER 1
1958

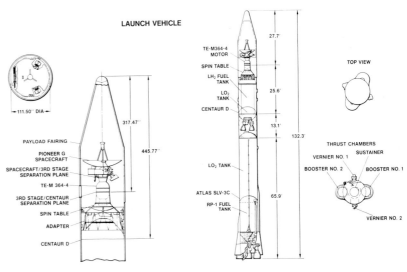

LAUNCH VEHICLE

111.50" DIA

317.47"

445.77"

PAYLOAD FAIRING

PIONEER G
SPACECRAFT

SPACECRAFT/3RD STAGE
SEPARATION PLANE

TE-M 364-4

3RD STAGE/CENTAUR
SEPARATION PLANE

SPIN TABLE

ADAPTER

CENTAUR D

TE-M364-4
MOTOR

SPIN TABLE

LH₂ FUEL
TANK

LO₂
TANK

CENTAUR D

LO₂ TANK

ATLAS SLV-3C

RP-1 FUEL
TANK

27.7'

25.6'

13.1'

132.3'

65.9'

TOP VIEW

THRUST CHAMBERS

VERNIER NO. 1   SUSTAINER

BOOSTER NO. 2   BOOSTER NO. 1

VERNIER NO. 2

e

During this same period an Outer Planets Panel associated with NASA's Lunar and Planetary Missions Board (also chaired by Van Allen) recommended in 1962 that planning should start for low-cost exploratory missions to the outer planets. Six years were to pass, however, before the Space Science Board of the National Academy of Sciences rated Jupiter as probably the most interesting planet from a physical point of view and recommended that "Jupiter missions be given high priority, and that two exploratory probes in the Pioneer class be launched in 1972 or 1973."

A mission to Jupiter originated at NASA Goddard Research Center, Greenbelt, Maryland in 1967 as a Galactic Jupiter Probe study for a project which was regarded as the natural logical progression in the United States' program of planetary exploration. The plan was to use the gravitational slingshot of Jupiter to explore far out into the Solar System and resolve some of the major questions troubling space physicists about interplanetary space at great distances from the Sun, and the magnetospheres and ionospheres of the outer planets.

In June 1969, a report issued by the Lunar and Planetary Missions Board emphasized the importance of finding out more about the outer giant planets and recommended that the Nation have a long-term plan to explore the outer Solar System.

Meanwhile, many proposals and scientific papers about exploration of the outer planets had been generated, including ideas for missions to visit several planets by one spacecraft using a slingshot effect derived from passing close to some of the planets.

NASA gave official approval for a mission to Jupiter in February 1969, assigning the program to the Planetary Programs Office, Office of Space Science and Applications, at NASA Headquarters, Washington, D.C. The Pioneer Project Office at Ames Research Center was appointed to manage the project, and TRW Systems Group, Redondo Beach, California, designed and built the spacecraft.

As a continuation of the Interplanetary Pioneer program, the mission to Jupiter was to consist of two identical spacecraft—referred to before launch as Pioneer F and Pioneer G—each weighing about 550 pounds and carrying 60 pounds of instruments for 13 scientific experiments. Each spacecraft was to be launched by an Atlas-Centaur.

For a few weeks every thirteen months the relative positions of Earth and Jupiter permit a spacecraft to be launched into a Jupiter-bound trajectory with minimum launch energy i.e., requiring the minimum amount of propellant for a given payload weight. Launch energies for the remainder of the 13 months are prohibitive. The first feasible opportunity for the Pioneer mission, considering the time needed to build the spacecraft, select its scientific experiments, and build instruments to perform these experiments, was in 1972, extending from late February through early March.

The initial objectives of the Pioneer mission to the giant outer planets as defined by NASA were to:

—Explore the interplanetary medium beyond the orbit of Mars.
—Investigate the nature of the asteroid belt from the scientific standpoint and assess the belt's possible hazard to missions to the outer planets.
—Explore the environment of Jupiter.

Initially the plans called for the spacecraft to be able to obtain about 12 spin-scan images of Jupiter.

To propel the 250-kgm (550-lb) spacecraft to the unprecedented velocity needed to enter a transfer trajectory to Jupiter, the Atlas-Centaur launch vehicle (figure 2-1e) was equipped with an additional solid-propellant stage.

Scientific experiments were selected over a series of planning meetings in the late 1960s, and by early 1970 they had all been decided upon.

At the May 1970 science conference, Glenn A Reiff, Pioneer Program Manager at NASA Headquarters, stated: "The mission is very exploratory and also very ambitious. The spacecraft will be injected into their trajectories at 50,000 ft/sec, the fastest that man has ever shot anything. The time of flight for each spacecraft is longer than for any other interplanetary flight so far, and the communications distances are greater. The spacecraft design itself is relatively new. We are including radioisotope thermoelectric generators as power supplies instead of solar cells." The spacecraft were to be spin-stabilized and would each carry a 9-foot-diameter antenna that would point toward Earth.

Charles Hall, Project Manager for Pioneer, stated that the opportunity for the first launch had to be within a very narrow time period. "It is about 17 or 18 days stretching from February 29 to March 17." He pointed out the mandatory nature of maintaining a very tight schedule if the first spacecraft and all its instruments were to be available for the launch window. The plan was to launch Pioneer F in late February or early March 1972 so it could arrive at Jupiter in November 1974, and to launch Pioneer G in April 1973 so it could arrive at Jupiter in December 1974.

Scientists were told that 13 scientific experiments would broadly study a number of intriguing interplanetary phenomena— the Sun's influence on interplanetary space, the configuration of the solar wind, and the penetration of galactic cosmic radiation into the Solar System—as well as make observations of Jupiter and its big satellites. One exciting task, should the spacecraft survive their encounter with the giant, was to search for the boundary of the heliosphere where galactic space begins and our Solar System ends.

Both spacecraft would spend at least 6 months passing through the asteroid belt. This region of space debris circling the Sun between the orbits of Mars and Jupiter might consist of parts of a planet that was unable to form because of gravity perturbations by the mighty Jupiter, or the remains of a planet that broke up. The debris was thought to range in size from 635-mile-diameter Ceres to specks of dust. At that time in 1970 there were widely varying esti-

mates of the number of asteroids smaller than one mile in diameter. Some estimates stated that there was enough material in the belt to form a small planet, together with a dangerously high number of small particles. Computers calculated the positions of the several thousand known asteroids to choose suitable flight paths for the two Pioneer spacecraft. However, the question of whether or not a spacecraft could safely pass through the belt was still debated hotly in scientific circles. There were many optimists, and Glenn Reiff stated: "We believe the chances are excellent of getting through the asteroid belt, but if hit by one the size of a marble the spacecraft would be put out of commission. Such a hit is thought to be an extremely remote possibility."

Because of its great distance from Earth, Jupiter is always seen almost completely illuminated by sunlight, so that little of the dark hemisphere of the giant planet can be observed. Earth-based observations had indicated that Jupiter radiates more energy than it absorbs from the Sun. Such a condition would indicate that the planet has a very dynamic interior with processes at work generating energy within it. The Pioneer spacecraft would provide an opportunity to look at the dark side of Jupiter and measure the planet's temperature in the shadow. An infrared radiometer carried by the spacecraft would allow scientists to analyze the thermal balance of Jupiter and establish whether or not the planet has an internal source of energy.

Another important task was to determine the intensity of energetic particles in the radiation belts which had been deduced from Earth-based radio observations of Jupiter. If these belts proved to be very extensive and filled with highly energetic particles, they would make it impossible for spacecraft to travel safely close enough to Jupiter to use its gravitational slingshot effect for voyages to the more distant planets, as envisioned for the Grand Tour.

At the May 1970 meeting, Van Allen commented. "The best evidence [for radiation belts of Jupiter] is really the radio astronomical evidence at the present time. It speaks quite clearly for the presence of very large numbers of energetic electrons trapped in the magnetic field of Jupiter. It [Jupiter's radiation environment] has much greater strength and much larger dimensions than Earth's radiation belts. We will measure the intensity of electrons at each point in the path of the spacecraft as it passes through.''

An instrument known as an imaging photopolarimeter would record images in red and blue light which would be used to build up images of the planet by scanning it in contiguous strips. The instrument would carry red and blue filters to separate the colors. Scientists hoped to be able to reconstruct a color picture of the planet. They also expected the images to show greater detail than can be seen from Earth even with the best telescopes.

At the meeting, Thomas Gehrels, University of Arizona, principal investigator for the photopolarimeter, told the assembled scientists that Earth-based observations indicated that the atmosphere of Jupiter over its poles is clear of the cloud bands that are so conspicuous in regions closer to the equator. He drew attention to this opportunity for the Pioneer spacecraft to look down onto the polar regions of the giant planet and perhaps see below the clouds that cover other regions so densely. We would be looking deep within an 86,000-mile-diameter ball of hydrogen flavored with helium, methane, ammonia, and water.

After its encounter with Jupiter, Pioneer F would fly out of the Solar System and become the first man-made object to travel to the stars. Pioneer G, which would be aimed to look into the polar regions of Jupiter, would then fly across the Solar System high above the plane of the planetary orbits. Commented Van Allen, "So far spacecraft have traveled only a few degrees outside the ecliptic plane. We want to know where the lines of force from the Sun return to it. This will be the first opportunity to investigate the magnetic field of the Sun above the ecliptic plane."

The two Pioneer Jupiter spacecraft were identical. The first, which would become Pioneer 10 after a

successful launch, explored the trial. Pioneer 11 would be launched so as to follow the path of Pioneer 10 thirteen weeks later, at the next launch opportunity. If the environment of the asteroid belt or of Jupiter would cause Pioneer 10 to fail in its mission, the second spacecraft, Pioneer 11, would provide a backup.

The Pioneer spacecraft were designed to operate reliably in space for many years. Each carried a data system to monitor the scientific instruments and to transmit over the vast distances to Earth scientific and engineering information about the condition of the spacecraft and its instruments. The spacecraft also had to be capable of receiving commands from Earth in order to perform their tasks and to change the operating modes of equipment aboard them as required. This would enable them to gather as much scientific information as possible about the unknown environment beyond Mars and the giant world of Jupiter with its retinue of bizarre satellites.

Each spacecraft's long curved path to Jupiter would stretch about 620 million miles around the Sun between the orbits of Earth and Jupiter. In the time it would take the spacecraft to reach Jupiter, Earth would travel almost twice around the Sun while Jupiter would move only about one sixth of the way around its orbit.

In-flight maneuvers were planned for several times during the mission to target the spacecraft to arrive at Jupiter at a time and position suitable for best observing the planet and several of its large satellites. The exact location of Jupiter and its satellites at any instant was not known precisely. The task of the first Pioneer would be to establish these positions (the ephemerides) more accurately so that Pioneer 11 could be sent closer to Jupiter and its big satellites.

Some tentative dates for the spacecraft's arrival at Jupiter would not be suitable because sensors would have been unable to perform the desired scientific experiments; other dates were unsuitable because they would have clashed with the arrival of another spacecraft, Mariner 10, at Venus or Mercury and given rise to conflicting requirements for the use of the big 210-foot antennas of the Deep Space Network needed to communicate with the Pioneers during the flybys. Yet other arrival dates were rendered unsuitable because of the relative motions of the planets. The launch vehicle boosted each spacecraft directly on to its trajectory to Jupiter, i.e., with no parking orbit around Earth, to start the flight to Jupiter at about 32,000 mph. A trip of just under 600 days was the shortest time to Jupiter within the capabilities of the launch vehicle, and a trip of 748 days, the longest. Approximately 715 days after launch, the motions of Earth and the spacecraft would position them on opposite sides of the Sun. Jupiter would appear very close to the Sun as observed from Earth. Arrival of Pioneer at Jupiter had to be timed to avoid this period because with Jupiter and the spacecraft appearing so close to the Sun as observed from Earth radio communications would be interrupted. Arrival at Jupiter beyond 700 days after launch was therefore unsuitable from a communications standpoint. For the launch in late February or early March 1972 a path was chosen so that the first Pioneer would arrive at Jupiter in early December 1973.

There were also decisions to be made as to how the spacecraft should approach Jupiter—how the trajectory should be inclined to the equatorial plane of Jupiter, and where the position of the closest approach should be located in the Jovian system.

It was quickly decided that the encounter trajectory (figure 2-2) for Pioneer 10 should be selected to give most information about the radiation environment, even if this resulted in damage to the spacecraft and caused the mission to end at Jupiter. The approach trajectory selected permitted a good view of the sunlit side of the planet as the spacecraft approached Jupiter with an opportunity, if the spacecraft survived its passage through the radiation belts, of seeing a partially illuminated (crescent) planet after the encounter. An occultation in which the spacecraft would pass behind Jupiter as viewed from Earth was arranged to gather information about the atmosphere of Jupiter unobtainable in any other way.

POSITION OF PIONEERS 10 AND 11 SINCE LAUNCH AND THE ORBITS OF EARTH, MARS, JUPITER, SATURN, AND URANUS

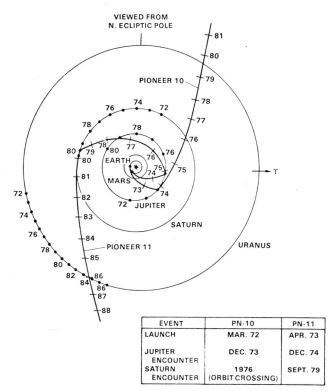

EVENT	PN-10	PN-11
LAUNCH	MAR. 72	APR. 73
JUPITER ENCOUNTER	DEC. 73	DEC. 74
SATURN ENCOUNTER	1976 (ORBIT CROSSING)	SEPT. 79

Figure 2-2: Trajectories of Pioneers 10 and 11 from Earth to Jupiter and outwards from the Solar System. The figures along the orbits and trajectories show the positions of the spacecraft and planets for the years stated. (NASA/Ames)

Then a decision had to be made on how close the spacecraft should approach Jupiter during its flyby. A workshop was held at the Jet Propulsion Laboratory in July 1971 at which scientists defined the environment of Jupiter as then understood, and this became the design environment for the spacecraft and its scientific instruments. There was a trade-off of conditions for the flyby. A close approach to Jupiter would increase the intensity of radiation, but the spacecraft would pass through the radiation more quickly and receive less total radiation. It was decided to send the first spacecraft to a close approach—only three times the radius of the planet, or twice the radius of Jupiter above the cloud tops. The consensus was that this was the closest the

PIONEER/JUPITER MAJOR SUBSYSTEMS

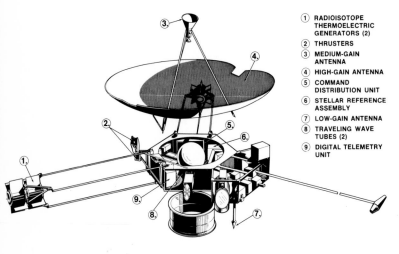

① RADIOISOTOPE
   THERMOELECTRIC
   GENERATORS (2)
② THRUSTERS
③ MEDIUM-GAIN
   ANTENNA
④ HIGH-GAIN ANTENNA
⑤ COMMAND
   DISTRIBUTION UNIT
⑥ STELLAR REFERENCE
   ASSEMBLY
⑦ LOW-GAIN ANTENNA
⑧ TRAVELING WAVE
   TUBES (2)
⑨ DIGITAL TELEMETRY
   UNIT

Figure 2-3: Diagram of the Pioneer/Jupiter spacecraft to identify the major subsystems.   (NASA/Ames)

dioisotope thermoelectric generators which seemed more suitable for the mission.

The spacecraft had to be accelerated to a very high velocity to reach Jupiter, and this limited the payload they could carry. Because on-board computing systems would be too heavy, the spacecraft had to be "flown from the ground," which meant that controllers would have to work around long delays in communications over the distance to Jupiter.

The duration of the mission demanded an unprecedented reliability on the part of all components of the spacecraft. This was achieved by reducing complexity in the spacecraft and by keeping the complex equipment on the ground. Vital items within the spacecraft, however, such as transmitters and receivers, were duplicated to provide a backup against failure.

A drawing of the spacecraft in figure 2-3 identifies its subsystems: general structure, attitude control and propulsion, communications, thermal control, electrical power, navigation, and payload of scientific instruments.

spacecraft could approach to avoid being seriously damaged by radiation. When the approach trajectory had been selected the scientific instruments could then be designed to survive the expected radiation intensities for this trajectory.

A pioneering mission to the giant planets faced the problem of obtaining electrical power for the spacecraft at the great distances where sunlight is much less intense than at Earth's orbit. Some early planning accepted solar cells because radioisotope power generators were not proven able to operate successfully for the years of operation required. Moreover, radiation from such generators was at that time not acceptable to the scientists since its effects on their instruments might invalidate data from the experiments. But very large solar arrays would be required and engineers expected that the radiation belts would damage the solar cells. Consequently, the project decided not to use solar cells but to rely on ra-

Structured around a 14-inch-deep, flat, hexagon-shaped equipment compartment, each spacecraft is 9.5 feet long. Attached to this hexagon a smaller, somewhat distorted hexagon compartment carries most of the scientific instruments. A 9-foot-diameter, parabolic, dish-shaped, high-gain antenna of aluminum honeycomb sandwich material is attached to the front of the equipment compartment (figure 2-4). A medium-gain antenna is mounted on three struts which project about 4 feet and carry the radio frequency feed for the big antenna. A low-gain, omnidirectional antenna extends 2.5 feet behind the equipment compartment.

Two trusses, each consisting of three rods, are mounted 120 degrees apart and project from two sides of the equipment compartment. Each supports at its end two radioisotope thermoelectric generators, held some 10 feet from the center of the spacecraft.

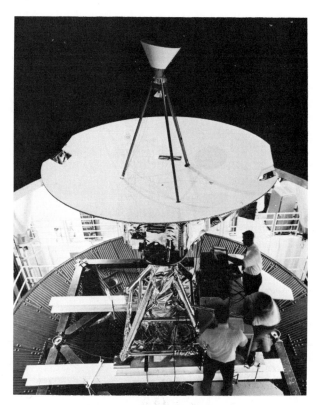

Figure 2-4: High-gain antenna used to beam data from the spacecraft over the half-billion miles to Earth. (TRW Systems)

A third single-rod boom, 120 degrees from the two trusses, projects from the experiment compartment to position a magnetometer sensor 21.5 feet from the center of the spacecraft. At launch the spacecraft fitted within the 10-foot protective shroud of the Atlas-Centaur launch vehicle, stowed there with its booms retracted and with its antenna dish facing forward, i.e., upward on the launch pad. All three booms were extended when the shroud had been jettisoned after launch.

Three pairs of rocket thrusters using a hydrazine propellant are located near the rim of the antenna dish (figure 2-5). They are used to direct the spin axis of each spacecraft, to keep it spinning at the desired rate of 4.8 revolutions per minute, and to change the spacecraft's velocity for in-flight maneu-

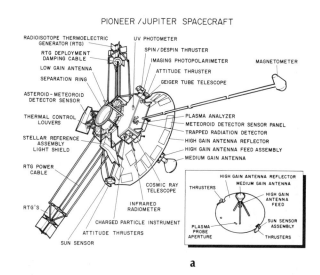

a

b

Figure 2-5: The Pioneer spacecraft, first to the outer planets of the Solar System.

(a) Diagram of the spacecraft showing its major systems and science experiments.

(b) Artist's concept of the spacecraft flying by Jupiter.

(NASA/Ames)

vers. These thrusters can be commanded to fire steadily or in pulses, and the firings can be precisely timed to the rotation of the spacecraft.

Except during flight near Earth and for periods when alignment has to be changed to make course corrections, the spin axis of each spacecraft, and hence the high-gain parabolic antenna, was kept pointed toward Earth.

To change the spacecraft's velocity the spin axis is first precessed, like the axis of a spinning gyroscope, by means of one pair of thrusters until it points in the direction along which the correcting velocity has to be applied. Then two other thruster nozzles, one on each side of the antenna dish, are fired continuously, both in the same direction—i.e., forward or aft, to apply the correcting velocity in the desired direction. To adjust the spin rate of the spacecraft, another pair is used. These are aligned tangentially to the antenna rim, one pointing against the direction of spin and the other pointing with it.

The spacecraft carries two identical receivers and two transmitters. The omnidirectional and medium-gain antennas operate together and are connected to one receiver, while the high-gain antenna is connected to the other. The receivers do not operate at the same time, but are interchanged by command, or, if there is a period of inactivity, they are switched automatically. Thus, if a receiver were to fail during the mission, the other would automatically start to operate in its stead. The radio transmitters, coupled to traveling-wave-tube power amplifiers, each produce 8 watts of transmitted power at the S-band of the radio frequency spectrum. Within this band, the communications frequency uplink from Earth to the spacecraft is at 2110 MHz, the downlink to Earth is at 2292 MHz.

Temperature within the spacecraft's instrument compartment is kept between −10° and 100° F, and at other levels elsewhere within the spacecraft. The temperature control subsystem is designed to adapt to less solar heat as the spacecraft moves into the outer Solar System, and to periods when the space-

craft passes through the shadow of a planet. Equipment compartments are insulated by multilayered blankets of aluminized plastic. Temperature-responsive louvers at the bottom of the equipment compartment control escape of excess heat. Other equipment has thermal insulation and electric heaters or one-watt radioisotope heaters.

Electric power for the spacecraft is derived from radioisotope thermoelectric generators which convert heat from the radioactive decay of plutonium-238 into electricity (figure 2-6). They are located on the opposite side of the spacecraft from the scientific instruments to reduce the effects of their neutron radiation. They developed a total of 155 watts at launch, but by the time the spacecraft reached Jupiter power output had decayed to about 140 watts from deterioration of the junctions of the thermocouples.

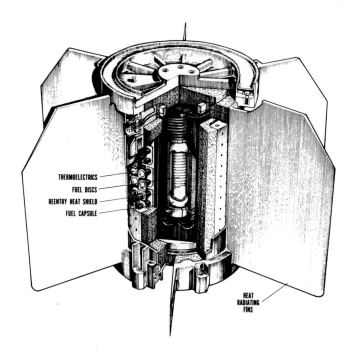

THERMOELECTRICS
FUEL DISCS
REENTRY HEAT SHIELD
FUEL CAPSULE

HEAT RADIATING FINS

SNAP 19/PIONEER RADIOISOTOPE THERMOELECTRIC GENERATOR

Figure 2-6: Radioisotope thermoelectric generator which provided electrical power for the spacecraft on its long journey to the limits of the Solar System.   (NASA/Ames)

Beyond Jupiter their output continues to decay but at a slower rate. Each spacecraft needs 100 watts to operate all its systems and experiments, of which one-quarter is consumed by the science instruments. Excess power is radiated into space as heat or sometimes used to charge a battery which supplies additional power when needed.

The shift in frequency of the radio signals received from the spacecraft is used to calculate the speed, distance, and direction of the spacecraft from Earth. Motion of the spacecraft away from Earth reduces the frequency of the spacecraft's radio signals. Called Doppler shift, this effect permits the speed of the spacecraft to be calculated by measuring the change in frequency of the received signals. An expected Doppler shift is compared with the observed shift to determine how closely the spacecraft follows its precalculated path, which is continually updated from analysis of the radio data. When the spacecraft reached Jupiter, the difference between the Doppler shifts expected and those observed not only provided information to keep the trajectory updated but also was used to determine masses of Jupiter and its major satellites.

Interplanetary space to and beyond Jupiter was investigated to resolve unknowns about the magnetic field in interplanetary space, cosmic rays (parts of atoms hurtling at enormous speed from the Sun and from the Galaxy), the solar wind (a blizzard of charged particles from the Sun) and its relationships with the interplanetary magnetic field, and interplanetary dust concentrations in the asteroid belt.

At Jupiter the spacecraft investigated the Jovian system in three main ways: measurement of particles, fields, and radiation; spin-scan imaging to provide images of the planet and some of its satellites; and additionally, by accurate observation of the path of the spacecraft, measurement of the forces—the gravity of the planet and its major satellites—acting upon it.

To achieve these scientific objectives, the spacecraft carried 11 scientific experiments which were selected from over 150 proposals submitted to NASA Headquarters in response to the official announcement of the opportunity to participate in the Pioneer mission. Later, a high-field magnetometer was added for the second spacecraft so that it would be able to measure more intense magnetic field strengths encountered by Pioneer 11 when it was sent closer to Jupiter and deeper into the radiation belts. In addition there were radio science experiments which did not require instruments to be carried by the spacecraft but made use of the radio transmissions from the spacecraft. The principal investigators for the experiments are listed in table 2-1.

Table 2-1   Pioneer Science Experiments

Experiment	Investigator	Affilliation
Magnetic Fields	E.J. Smith	Jet Propulsion Laboratory
Fluxgate Magnetometer	M.H. Acuna	Goddard Space Flight Center
Plasma Analyzer	J.H. Wolfe	Ames Research Center
Charged Particle Composition	J.A. Simpson	University of Chicago
Cosmic Ray Energy Spectra	F.B. McDonald	Goddard Space Flight Center
Jovian Charged Particles	J.A. Van Allen	University of Iowa
Jovian Trapped Radiation	R.W. Fillius	University of California, San Diego
Asteroid-Meteoroid Astronomy	R.K. Soberman	General Electric Company
Meteoroid Detection	W.H. Kinard	Langley Research Center
Ultraviolet Photometry	D.L. Judge	University of Southern California
Imaging Photopolarimetry	T. Gehrels	University of Arizona
Infrared Thermal Structure	G. Munch	California Institute of Technology
Celestial mechanics	J.D. Anderson	Jet Propulsion Laboratory
S-Band Occultation	A.J. Kliore	Jet Propulsion Laboratory
Program Scientist	A.G. Opp	NASA Headquarters
Project Scientist	J.H. Wolfe	Ames Research Center

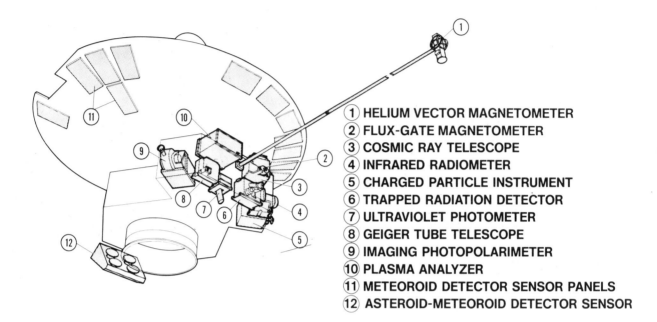

1. HELIUM VECTOR MAGNETOMETER
2. FLUX-GATE MAGNETOMETER
3. COSMIC RAY TELESCOPE
4. INFRARED RADIOMETER
5. CHARGED PARTICLE INSTRUMENT
6. TRAPPED RADIATION DETECTOR
7. ULTRAVIOLET PHOTOMETER
8. GEIGER TUBE TELESCOPE
9. IMAGING PHOTOPOLARIMETER
10. PLASMA ANALYZER
11. METEOROID DETECTOR SENSOR PANELS
12. ASTEROID-METEOROID DETECTOR SENSOR

Figure 2-7: Location of the science instruments.
(NASA/Ames)

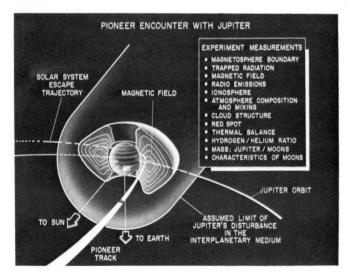

Figure 2-8: Summary of the experiments of Pioneer as it encountered Jupiter and passed through the anticipated magnetosphere of the giant planet. (NASA/Ames)

Locations of the scientific instruments on the spacecraft are shown in figure 2-7. Some of the experiments at Jupiter are summarized in figure 2-8.

Particles and Fields Experiments

Magnetic fields interact with the plasma of electrically charged particles in interplanetary space and control its flow from the Sun across the Solar System. Before the Pioneer mission to Jupiter scientists were uncertain about how the magnetic field of the Sun controlled the flow of plasma beyond Mars. The outer boundaries of this influence were vague, and how the plasma and fields of the Solar System interact with those of the Galactic System were completely unknown. The two Pioneers will continue to explore outward and may find the transition region of the solar influence, which is called the heliopause.

Pioneer carried a sensitive helium vector magnetometer to measure the fine structure of the interplanetary field, map the field of Jupiter, and provide field measurements to evaluate the interaction of the solar wind with the planet. The magnetometer is sensitive to fields from 10^{-7} to 1.4 gauss. (The surface field of the Earth is approximately 0.5 gauss; 1 gauss being the unit of magnetic flux density.)

At the heart of the instrument a cell filled with helium is excited by electrical discharge at radio frequencies and by infrared optical pumping. An infrared detector measures changes in helium absorption caused when a magnetic field passes through the instrument.

Pioneer 11 carries an additional magnetic field measuring instrument, a fluxgate magnetometer to measure the intense field closer to Jupiter. The instrument has two magnetic ring cores. These are driven to saturation at a frequency of 8 kHz. An external magnetic field causes an imbalance in the sensors which is detected by four coil windings. This instrument can measure a field up to 10 gauss compared with the 1.4 gauss of the other magnetometer.

How the solar wind behaves at great distances from the Sun was highly conjectural before the flight of the first spacecraft to Jupiter, and how the solar wind interacted with the giant planet was almost completely unknown. To resolve the question, a plasma analyzer looks toward the Sun through a hole in the spacecraft's big dish-shaped antenna. Solar wind particles enter the instrument between two quadraspherical plates where the direction of arrival, the energy, and the number of ions and electrons of the solar wind are measured (figure 2-9).

Analyzers provide high and medium resolution so that the instrument can detect particles of different energy levels. A high resolution analyzer measures the number of ions per second between 100 and 8,000 eV (electron volts). A medium resolution analyzer counts ions with energies from 100 to 18,000 eV, and electrons from 1 to 500 eV.

A charged particle detector has two particle telescopes to operate in interplanetary space, and two to measure trapped electrons and protons within the radiation belts of Jupiter. In interplanetary space this experiment looks for hydrogen, helium, lithium, boron, carbon, and nitrogen, and for deuterium, he-

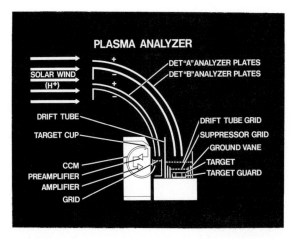

Figure 2-9: Diagram of the plasma analyzer instrument used to measure high energy particles in the solar wind and the magnetosphere of the planet.   (NASA/Ames)

lium-3, and helium-4 to differentiate between particles coming from the Sun and those coming from the Galaxy. The main telescope measures the composition of cosmic rays from 1 to 500 MeV (million electron volts), and a low-energy telescope measures 0.4 to 10 MeV protons and helium nuclei.

For the environment of Jupiter two new types of sensors were developed. A solid-state electron current detector, operating below $-104°$ F, detects electrons above 3.3 MeV which generate decimetric radiation from Jupiter. A trapped proton detector is insensitive to electrons but detects protons possessing energies above 35 MeV.

To obtain the energy spectra of cosmic rays the Pioneers each carried a cosmic ray telescope. This experiment also monitors solar and galactic cosmic rays and tracks high-energy particles from the Sun. The instrument has three, three-element, solid-state telescopes to measure the flux of protons between 56 and 800 MeV, protons with energies between 3 and 22 MeV, and the flux of electrons between 0.05 and 1 MeV and of protons between 0.05 and 20 MeV. It can also identify the nuclei of the ten lightest elements (hydrogen to neon).

A charged particles instrument measures the intensities, energy spectra, and angular distribution of energetic electrons and protons in interplanetary space and in the environments of Jupiter and Saturn. It consists of an array of seven miniature Geiger-Muller tubes, each a small gas-filled cylinder. When a charged particle passes through the gas, an electrical pulse is generated by the applied voltage. Protons of energy greater than 5 MeV and electrons of energy greater than 40 kev are detected. For Pioneer 11 one Geiger-Muller tube was replaced by a thin silicon wafer, so that protons in the energy range 0.61 to 3.41 MeV could be detected.

Using other types of telescopes a trapped radiation detector covers a broader range of energies of electrons and protons. An unfocused Cerenkov counter, which detects light emitted as particles pass through it, records electrons of energy 0.5 to 12 MeV. An electron scatter detector records electrons with energies between 100 and 400 keV. The instrument also has a minimum ionizing detector to detect ionizing particles of less than 3 MeV and protons between 50 and 350 MeV. Two scintillation detectors distinguish between electrons of less than 5 keV and protons of less than 50 keV.

Meteoroids and Interplanetary Dust Experiments

Two experiments investigate particles in interplanetary space; one detects light reflected from particles, the other detects impact of small particles. The reflected light detector consists of four optical devices to detect sunlight reflected from meteoroids passing through their fields of view. Each has an 8-degree field of view that overlaps slightly with the field of its neighbors. Whenever a particle appears simultaneously on any three of the telescopes it is recorded and its distance, trajectory, velocity, and relative size are calculated. This instrument can detect objects ranging from asteroids miles in diameter at great distances to sunlit dust particles a few feet from the spacecraft.

The other experiment employs 13 panels, each containing 18 sealed cells pressurized with argon and nitrogen mounted on the back of the main antenna dish. When a cell is punctured by a small interplanetary particle, its loss of internal pressure is detected. The experiment can detect meteoroids as small as one billionth of a gram. On Pioneer 11 the panels were thickened so that particles of greater mass could be detected.

Optical Experiments

Images of Jupiter and its satellites were obtained by an imaging photopolarimeter (figure 2-10) which operates in three modes, differing mainly in their sensitivity and field of view. The most sensitive was used during the interplanetary journey to measure zodiacal light, Gegenschein (counterglow), and integrated starlight from the Galaxy. The other modes gathered photometric and polarimetric data on Jupiter over a wide range of phase angles and distances, to determine the shape, size, and refractive index of cloud particles and the abundance of gas above the clouds. Information was also obtained about the Galilean satellites. Data were computer processed into images of Jupiter and its large satellites.

The imaging experiment relies upon the spin of the spacecraft to sweep the field of view of a small telescope across a target (planet, satellite, or distant space) in 0.003-degree strips, recording either red or blue light. These strips are later placed side by side by a computer to form a visual image.

An ultraviolet photometer measures reflected or emitted ultraviolet light and was used to investigate interplanetary hydrogen, helium, and dust, and the environment of Jupiter and some of its satellites. Scattering of the Sun's ultraviolet light in space by these gases is measured by this instrument. The Solar System is traveling through an interstellar gas cloud of cold, neutral (uncharged) hydrogen. The neutral hydrogen could result from neutralization of fast solar wind hydrogen ions at the boundary of the he-

PHOTOPOLARIMETRY

CHANNELTRON
DUAL CHANNEL

WOLLASTON PRISM

f/3.4 FOCAL PLANE OPTICS:
FIELD-OF-VIEW APERTURES
DEPOLARIZERS, RETARDER,

COMPENSATOR, CALIBRATION SOURCE

ENTRANCE PUPIL

FOLDING MIRROR

FIELD LENS

RELAY LENS
SPECTRAL FILTERING
COATING, RED
TRANSMITTING

FOLDING MIRROR

DICHROMATIC MIRROR

RELAY LENS
SPECTRAL FILTERING
COATING, BLUE
TRANSMITTING

CALIBRATION LAMP

FOLDING MIRROR

TELESCOPE OPTICS

a

b

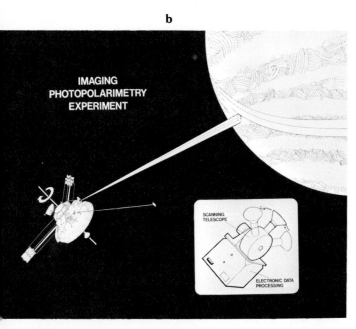

IMAGING
PHOTOPOLARIMETRY
EXPERIMENT

SCANNING
TELESCOPE

ELECTRONIC DATA
PROCESSING

Figure 2-10: The photopolarimeter instrument which obtained the first images of Jupiter and Saturn from a spacecraft.

(a) Diagram of the instrument. (NASA/Ames)

(b) How the spinning of the spacecraft swept the viewpoint of the photopolarimeter across Jupiter to form images of the planet. (NASA/Ames)

liosphere. Converted into fast uncharged hydrogen atoms, they would be expected to diffuse back into the heliosphere. An alternative source might be the Galaxy, with the hydrogen pushed into the Solar System by the system's 45,000 mph motion through interstellar space.

The viewing angle of the photometer is fixed. The spin of the spacecraft scans the instrument around the celestial sphere. At Jupiter the photometer scanned above the cloud tops searching for hydrogen and helium. Within the Jovian system the instrument measured the scattering of solar ultraviolet light by the atmosphere to obtain information about the amount of atomic hydrogen in the upper atmosphere, the mixing rate within the atmosphere, the amount of helium, and the ratio of helium to hydrogen. This ratio is important in order to resolve different theories of origin of the giant outer planets and their subsequent evolution. Before Pioneer's mission helium had not been identified in the Jovian atmosphere though it was believed to be there. The instrument also sought an ultraviolet glow that would indicate the presence of polar auroras on Jupiter.

Emissions of infrared radiation from Jupiter can be measured from Earth and infrared maps of the planet show belts and bands similar to those seen in visible light. But most of the planet's infrared radiation is emitted at 20 to 40 micrometers, wavelengths blocked by the Earth's atmosphere. A two-channel infrared radiometer carried by Pioneer measured radiation at 14 to 25 and 25 to 56 micrometer wavelengths to provide more accurate measurements of Jupiter's net heat energy output, the temperature across the planet's disc, and information to help ascertain the thermal structure and chemical composition of the atmosphere.

As with the ultraviolet instrument, the radiometer is fixed and relies upon the spin of the spacecraft to scan the planet's cloud tops. It has a 3-inch aperture Cassegrain optical system, and uses thin-film, bimetallic thermopiles to detect infrared radiation.

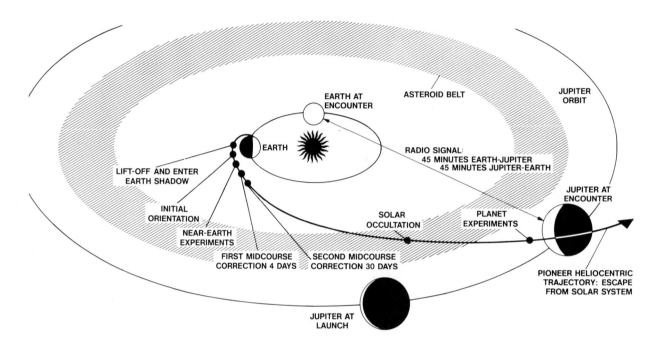

Figure 2-11: The mission through the asteroid belt to Jupiter.
(NASA/Ames)

Celestial Mechanics Experiments

In addition to experiments that rely upon specific instruments on the spacecraft, some experiments use the spacecraft itself and its radio signals. Tracking the spacecraft provides its velocity along the line from Earth to the spacecraft to within a fraction of a millimeter per second. The Doppler tracking data have been used to refine the calculated value of the masses of Jupiter and the Galilean satellites—a five-fold improvement in the calculation of Jupiter's mass and masses of the satellites to an accuracy of better than one percent. The experiment also determined the polar flattening of the planet to within one-half mile.

Radio signals from Pioneer were also used to probe the atmosphere of Jupiter and its innermost large satellite, Io. As the S-band radio signal passed through the Jovian atmosphere when Pioneer moved behind Jupiter its changed characteristics provided information on the ionosphere and the atmosphere's

density to a pressure level of about one Earth's atmosphere. As the spacecraft passed behind Io, the experiment sought for evidence of an atmosphere of that satellite.

First Spacecraft Launched to Jupiter

Trajectory analysts evaluated approaches to Jupiter (figure 2-11). Targeting so that the spacecraft would be occulted by a satellite during the encounter would allow the radio signals to probe through any atmosphere possessed by that satellite and provide information from which scientists could deduce its composition. Io was a prime candidate because it modulated some of the radio emissions from Jupiter.

From the time the contract to build the spacecraft was awarded to the scheduled launch date, early March 1972, was slightly less than 2½ years (figure 2-12). But the thousands of people involved in the Pioneer program met this tight schedule—one more example of how space program people can rise to difficult tasks and achieve the "impossible" when provided with a stimulating and creative challenge.

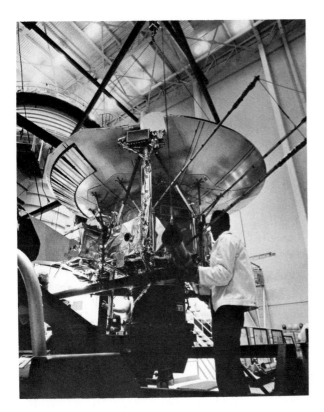

Figure 2-12: Engineering model of the Pioneer/Jupiter spacecraft is readied for thermal testing in TRW's 22-by-46-foot space simulation chamber at Redondo Beach, California. The 9-foot antenna, tilted toward the ceiling in this photo, was directed toward Earth during the mission.   (TRW Systems)

Project manager Charles F. Hall praised them: "It is most appropriate to compliment the many dedicated people who have worked so hard to reach this first goal of the Pioneer . . . mission to Jupiter and to congratulate all for a job well done. I estimate that at the time of the Pioneer [10] launch, more than 15 million man-hours will have been expended to make this goal possible. I am sure that you all feel as I do that a successful mission wherein we will be exploring new frontiers in space will be a just compensation for this large effort and that we are, indeed, a fortunate, select group which has been given the opportunity to participate in and contribute to the Pioneer . . . mission."

Pioneer proved, indeed, to be a further example of how, when provided with resources through government, dedicated people can enterprisingly get things done despite the procrastination of bureaucracy. In common with nearly all other space programs, Pioneer faced funding problems as government support waxed and waned. Space project managers have stated that one of the most difficult parts of their job is adjusting to these peaks and valleys of funding while keeping the dedicated teams together and trying to meet the inexorable deadlines of planets moving along their orbits.

On December 22, 1971, an Atlas/Centaur stood on Launch Complex 36A at the John F. Kennedy Space Center, Florida, awaiting the Pioneer. The spacecraft, equipped with its full complement of scientific instruments, traveled by air on January 14, 1972, from TRW Systems, California, to the launch site. There it was thoroughly tested (figure 2-13) and its radioisotope thermoelectric generators were installed and propellant was loaded. Finally, it was mated to the third stage of the launch vehicle, encapsulated in its protective nose-shroud, and mated with the Atlas/Centaur on the launch pad.

When the launch window opened on February 27, 1972, blockhouse electrical power failed within an hour of the planned liftoff. By the time this had been corrected high winds made it too hazardous to launch the spacecraft. These winds continued, and it was not until March 2, 1972, at 8:49 P.M. EST, that the Atlas/Centaur lifted from the launch pad (figure 2-14) carrying Earth's first space probe to Jupiter and beyond. The launch vehicle rose majestically into the night sky, its brilliant exhaust dwarfing the lightning flashes, its mighty roar drowning the rolls of distant thunder.

During prelaunch activities and the launch itself, the spacecraft and launch vehicle were controlled at the Kennedy Space Center by launch teams from the Ames and Lewis Research Centers, respectively. Telemetered data poured into the control center from the launch vehicle and from the spacecraft. Pioneer withstood the pounding thrust of the booster's

Figure 2-13: Pioneer is here shown being readied for launch in the clean room at the Kennedy Space Center. Its launch fairing is at the left.   (NASA)

Figure 2-14: The first spacecraft headed for Jupiter leaves the launch pad at Cape Canaveral, Florida, March 2, 1972.

(NASA)

mighty rocket engines as the spacecraft was given enough kinetic energy to break free from Earth's gravitational fetters at 32,114 mph. After separating from the launch vehicle the spacecraft deployed the booms of its radioisotope power units and that of the magnetometer.

Shortly after the spacecraft separated from the Atlas-Centaur and entered the transfer orbit to Jupiter, control of the spacecraft was transferred to a cadre of personnel forming the Ames flight operations team at the Jet Propulsion Laboratory, and control of the scientific instruments within the spacecraft was transferred to the Pioneer Mission Operations Center (PMOC) at NASA's Ames Research Center. Engineering specialists monitored all the spacecraft's subsystems, such as telemetry, power, thermal, at-

titude control, data handling, and command. Quick reaction to unusual events was mandatory at this time when the spacecraft had been subjected to the stresses of launch.

Launching was most accurate. The spacecraft's velocity required a correction on March 7 of 31.3 mph only. The correction was made so that the time of arrival at Jupiter would be more suitable for some of the experiments.

The magnetometer and charged particle detectors were turned on to obtain in-flight calibration in the well-known magnetic and radiation environment around Earth. Two days after liftoff, the cosmic ray telescope was turned on, and then the ultraviolet photometer, the asteroid-meteoroid detector, imag-

NAVIGATION

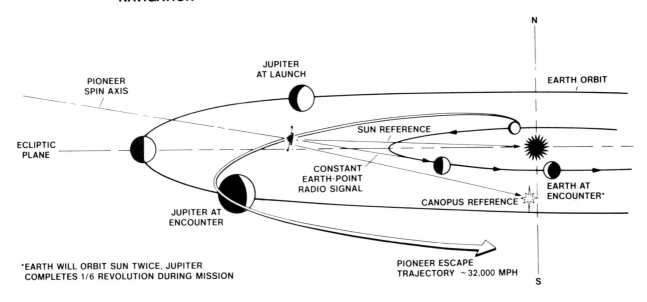

PIONEER SPIN AXIS

JUPITER AT LAUNCH

EARTH ORBIT

ECLIPTIC PLANE

SUN REFERENCE

CONSTANT EARTH-POINT RADIO SIGNAL

CANOPUS REFERENCE

EARTH AT ENCOUNTER*

JUPITER AT ENCOUNTER

N

S

*EARTH WILL ORBIT SUN TWICE. JUPITER COMPLETES 1/6 REVOLUTION DURING MISSION

PIONEER ESCAPE TRAJECTORY ~32,000 MPH

Figure 2-15: Pioneer 10 soon passed the orbit of Mars and entered the unexplored region of the asteroid belt which had to be crossed safely for a successful mission to the giant planets. The spin axis of the spacecraft was kept pointing toward Earth and navigators used references on the star Canopus and the Sun to guide the spacecraft to its distant destination. (NASA/Ames)

ing photopolarimeter, and the plasma analyzer. All were on and operating satisfactorily by 10 days after launch.

When the spacecraft was near Earth, sunlight illuminated it from the side and heated the spacecraft. It was commanded to point its spin axis so that the antenna's shadow shielded vulnerable parts from the Sun. The spacecraft was later oriented for its long voyage so that the big dish-shaped antenna pointed toward Earth.

Several days after liftoff, following some maneuvers of the spacecraft to adjust its trajectory, and with all equipment and science instruments performing well, the mission crews left the Jet Propulsion Laboratory and the John F. Kennedy Space Center and returned to Ames Research Center, the control hub for the rest of the mission.

Performance of Pioneer 10 continued to be excellent. The Pioneer spacecraft system manager, Ralph W. Holtzclaw, stated: "Now that we have had a chance to recover from the emotional trauma of getting Pioneer 10 launched, it is time to sit down and perform a factual engineering examination of this machine 'we' have wrought. As Pioneer 10 settles into the 'cruise' phase of its voyage to Jupiter, many analyses must be made of the live operation of this vehicle in a space environment—to ensure specified performance during the crucial Jupiter encounter. Preliminary indications are that Pioneer 10 is a good spacecraft and a good mission."

Once the spacecraft settled down to the interplanetary mode, spacecraft events occurred more slowly. The task changed to one of watching and waiting and becoming familiar with an increasing delay for signals to go to the spacecraft and return to Earth. In this interplanetary "cruise," all monitoring of the process of "flying" the spacecraft was by a small group at the Pioneer Mission Operations Center. The spacecraft was navigated (figure 2-15) with reference to the Sun and the star Canopus, with the spin axis of the spacecraft maintained so that the big dish antenna pointed toward Earth.

During Pioneer's interplanetary journey engineering and scientific data being returned from the spacecraft were continually monitored by computer and by people to detect problems at the earliest possible moment. The computer monitored telemetry signals. It generated an audible alarm and a printed message if it found anything was not as preplanned. Day or night, when required, the duty operator brought any such problem to the attention of a cognizant engineer or scientist who could resolve it. Specific procedures were provided to the trained mission controllers to cover any emergency, and to advise them whom to contact for a decision if unexpected problems surfaced.

During the long voyage through interplanetary space, data from each scientific instrument was sampled periodically to check that it was operating correctly. Some of the instruments began to gather important new data about the interplanetary environment.

Preliminary Interplanetary Discoveries

On a moonless night on Earth when skies are clear and there is no glare from artificial lighting, a faint glow is seen concentrated near to the path of the Sun through the constellations of the zodiac. In Northern Hemisphere spring, when the ecliptic is oriented at a large angle to the western horizon, the zodiacal light appears as a cone of light when the western sky has darkened after sunset; and in autumn, this cone appears in the east, before the rising of the Sun. The glow was understood to originate from sunlight reflected by countless numbers of small particles orbiting the Sun in the space between the planets.

Pioneer 10 measured the intensity of the zodiacal light in interplanetary space, and for the first time investigated, away from Earth, a concentration of the zodiacal light in a direction away from the Sun, called the Gegenschein or counterglow. Pioneer showed that the counterglow shines as far as Mars, thereby confirming that it is not related to the Earth

as a kind of cometary tail from our planet stretching away from the Sun. Instead it is caused by the particles which are responsible for the zodiacal light showing fully illuminated faces to Earth when directly opposite to the Sun, analogous to many tiny full moons.

On July 15, 1972, Pioneer 10 entered the asteroid belt. Project officials expected a safe passage, a 90 percent chance of passing through undamaged, but there was a risk that they could be wrong. Pioneer would not approach close to any known asteroid, but any particle exceeding $\frac{1}{50}$ of an inch in diameter could seriously damage the spacecraft. Such a particle might hit the spacecraft at 15 times the speed of a bullet from a high powered rifle.

In February 1973, Pioneer 10 emerged unscathed from the asteroid belt, at a distance of 340 million miles from the Sun. The belt contains much less material in small particle sizes than had been anticipated. Pioneer had surmounted the first obstacle to exploration of the outer Solar System.

Second Spacecraft Safely Launched

About this same time, Pioneer 11 was being readied for launch at the Kennedy Space Center. If Pioneer 10 should fail during the rest of its mission, Pioneer 11 would be commanded to repeat the failed part of the mission. Otherwise, Pioneer 11 would be retargeted to fly a different course through the Jovian environment to obtain another set of samples of that environment. If all conditions proved favorable, the path past Jupiter could be arranged so that the second Pioneer could continue across the Solar System to an encounter with Saturn almost five years later.

The first launch window for Pioneer 11 opened at 9:00 P.M. EST on April 5, 1973. The countdown was marred by a spell of bad weather with severe thunderstorms. But crews at the site worked strenuously to overcome the delays and made the spacecraft ready for the opening of the launch window. The

Atlas-Centaur roared into space at 9:11 P.M. EST on April 5 (figure 2-16). Pioneer 11 separated from the launch vehicle at 9:26 P.M. Then there was trouble. As the booms were deployed one of them did not extend to its full extent. The effect was to keep the spacecraft spinning too quickly.

Thrusters were fired in an attempt to vibrate the boom loose. It did extend slightly. Next the spacecraft was reoriented to reduce solar heating and, surprisingly, the boom extended fully of its own accord and all was well. The spacecraft then rotated on its spin axis at the correct rate of 4.8 rpm.

Pioneer 11 repeated the experiments of Pioneer 10 during the interplanetary cruise, and it, too, passed safely through the asteroid belt. Pioneer 11 was targeted to be closest to Jupiter at 9:21 P.M. PST on December 2, 1974, just about one year after Pioneer 10. Afterwards it would become the first spacecraft to fly by Saturn in September 1979.

Figure 2-16: On April 5, 1973, Pioneer 11 followed Pioneer 10 into the outer Solar System.   (NASA)

Important Interplanetary Discoveries

On its journey to the first encounter of a spacecraft with Jupiter, Pioneer 10 made important scientific observations in the unexplored space beyond the orbit of Mars. After encounter these observations continue until at least the mid-1980s, when Pioneer 10 will be near the orbit of Pluto. They are supplemented by information from Pioneer 11, the second spacecraft to reach Jupiter and the first to reach Saturn. Pioneer 11 was the first spacecraft to explore interplanetary space high above the plane of the ecliptic during its voyage from Jupiter to Saturn. It, too, continues to send back information from far beyond the orbit of Saturn. The Voyager spacecraft add considerable new data and may continue to send data from much greater distances than the Pioneers.

The result is that we now have a much improved understanding of the interplanetary space of our Solar System beyond Mars from the data gathered by these four record-breaking spacecraft. The new information describes the interplanetary medium far beyond the orbit of Mars, and to a height of 100 million miles above the plane of the ecliptic. It also describes the asteroid belt, how the solar wind blows through the outer Solar System, and how cosmic rays act as they plunge into the Solar System.

The Night Sky Glows Are Far Beyond Earth's Skies

As mentioned earlier, Pioneer 10 confirmed theories about the zodiacal light which attributed the faint glowing band of light along the zodiac to sunlight reflected from large numbers of particles in interplanetary space. the Gegenschein, or counterglow, which appears as an enhancement of the zodiacal light exactly opposite to the Sun in Earth's sky, was thought to be produced by reflection of sunlight from distant particles. One of several other theories was that the reflection came from a stream of particles somewhat like a comet's tail extending from the Earth away from the Sun.

The imaging photopolarimeter resolved the problem. During the first few weeks of the mission, observations concentrated on the counterglow and soon showed that it could not be associated with Earth. Although the spacecraft had not moved much farther from the Sun, it had moved around the Sun ahead of the Earth. Yet the direction of the counterglow seen from the spacecraft was still directly opposite the Sun, and now this direction was different from the direction to the counterglow as seen from Earth. Obviously the counterglow must originate from particles in the Solar System, and not be connected with the Earth.

The brightness of the counterglow decreases with increasing distance from the Sun although the glow is still present at the orbit of Mars. The glow is enhanced slightly within the asteroid belt, which shows that the reflecting particles increase in number slightly within the belt. However, it is clear that the particles causing the Gegenschein are concentrated in the inner Solar System, for beyond the asteroids there is virtually no counterglow.

Pioneer also mapped the zodiacal light and discovered that, except within the asteroid belt, this light decreases in brightness as the square of an observer's distance from the Sun. Like the Gegenschein, the particles responsible for the zodiacal light are concentrated in the inner Solar System, but are slightly more numerous within the asteroid belt. There is virtually no zodiacal light beyond 3.5 times Earth's distance from the Sun. Free from the obscuring effect of background zodiacal light, Pioneers' instruments measured integrated starlight from the Galaxy.

Meteoroids and Asteroids

At one time it had been speculated that because Soviet and U.S. spacecraft on their way to Mars had encountered trouble at about 110 million miles from the Sun there was a concentration of small asteroids associated with many dust particles inside the orbit of Mars. Press reports commented on the "Great Galactic Ghoul" lying in wait to gobble spacecraft from

Earth! Pioneer 10 showed the speculation to be unfounded. The data from the spacecraft proved that the opposite was true: Mars sweeps its orbit clean of particles.

Very small particles of interplanetary dust are swept from space by Earth as well as by Mars. These planets produce a gap in Solar System dust extending from 1.14 to 1.34 times Earth's distance from the Sun.

Beyond Mars, even the 175-million-mile-wide asteroid belt did not prove to be as hazardous as had been anticipated prior to the epic voyage of the Pioneers. The large number of minor planets discovered over the years in the asteroid belt suggested that collisions might be frequent, and in the billions of years since the Solar System formed, these collisions would have generated an enormous number of particles, ranging in size from tiny particles of dust to the major asteroids. Such particles would be a serious hazard to spacecraft.

By October 1972, when Pioneer 10 had traveled halfway through the belt, the counting rate for meteoroid impacts had remained much the same as it had been before the spacecraft entered the belt. Subsequently the rate was relatively constant all through the belt. There was no trace of myriad tiny bodies capable of damaging spacecraft in these regions of space. Consistent with the observations of zodiacal light and the counterglow, fine particles appear to be evenly distributed between the planets. Pioneer 11, when it safely passed through the asteroid belt, confirmed the findings of Pioneer 10 about the numbers of particles in the belt. As Pioneer 11 traveled from Earth's orbit, the number of the smallest particles its instruments could detect (which were about 0.001 mm in diameter) declined. However, the spacecraft found that particles between 0.01 mm and 0.1 mm in diameter are evenly distributed through the asteroid belt. But there are three times as many particles from 0.01 to 1.0 mm diameter in the center of the belt as there are near Earth.

Pioneer 11 discovered that between 112 and 214 million miles from the Sun large particles are almost absent. Its detector recorded only one penetration. But in the asteroid belt, there are larger particles again. Yet Pioneer 11 recorded only about one-sixth as many as did Pioneer 10. Analysis of the results on the basis of the different abilities of the two Pioneers to measure particles—Pioneer 11's cells had thicker walls—implies that particles of 0.01 to 0.1 mm size are three times as common as larger particles in the asteroid belt. Data from both Pioneers show almost three times as many large particles inside the asteroid belt as there are between Earth and the belt.

Nevertheless, the Pioneers demonstrated that the asteroid belt is not a region of large numbers of high-velocity projectiles capable of damaging spacecraft. The number of such particles was found to be far lower than had been predicted. Although the belt does contain very large bodies as well as tiny dust particles, the Pioneers proved that there are no concentrations of high-velocity dust particles in the belt to cause a hazard to spacecraft.

The Pioneers discovered that there are small particles concentrated in the vicinity of Jupiter by its gravity, but not to the point of being a hazard to spacecraft flying by the planet.

The Solar Wind

The solar wind as it moves through the interplanetary medium might be expected to expand symmetrically from the Sun and cool adiabatically. Its temperature would decrease with distance according to a four-thirds power law. Earlier spacecraft had shown that such a law is not valid at Earth's distance from the Sun. There are nonuniformities in the solar wind which arise from hot parts of the solar corona. The speed of the solar wind when it leaves the Sun is determined by the temperature of the corona from which the wind originates, and a hot part of the corona produces high-speed wind streams. Rotation of the Sun swings the streams in spirals though space and fast-moving streams catch up with slower streams that left the Sun earlier.

The result is an interaction of the magnetic fields carried by the solar wind which produces steep magnetic gradients. These magnetic gradients scatter cosmic rays coming from the Galaxy into the Solar System, and prevent low-energy cosmic rays from penetrating into the inner Solar System. Scientists faced the big question of how far low-energy galactic cosmic rays penetrate into the Solar System. This question is important because the cosmic rays carry information about the composition of the stars.

The Pioneers showed that to 2 billion miles from the Sun the strength of the Solar magnetic field, the density of the solar wind, and the numbers of high energy particles from the Sun, all decline approximately as the square of the distance from the Sun. At great distances from the Sun, the stream of the solar wind has fewer variations and its gases cool much less rapidly because the kinetic energy of high-speed streams is converted into thermal energy. Beyond Saturn the Sun's magnetic field is tightly wound into a spiral. But nevertheless the solar wind still carries a signature of events on the Sun. Energetic particles accelerated by solar flares reach the outer regions of the heliosphere. It often takes several weeks for the peak flux of particles to diffuse to great distances from the Sun. One surprising aspect was observed by Pioneer 10 beyond 15 times Earth's distance from the Sun. The proton flux very slowly rose to a maximum level and then remained constant for several rotations of the Sun. However, Earth-based data show no continual high flux of solar protons. To account for the Pioneer observations, the heliosphere near the ecliptic plane must be filled with a high flux of protons.

Several cases of local acceleration of particles by the outward-moving solar wind blasts were also observed. These were the first such observations in the outer heliosphere.

Cosmic Rays

The Pioneers continue to provide data about the radial and latitudinal intensity gradients of cosmic ray protons, helium, and heavier elements coming from the Galaxy. They discovered small but slightly variable increases in intensity with increasing distance from the Sun. Results to 1981 suggest that the boundary between the heliosphere and the interstellar medium is beyond 20 times Earth's distance from the Sun—farther than was thought before the two Pioneer spacecraft penetrated to the outer Solar System.

The Pioneer spacecraft provided observations of how increasing levels of solar activity affect propagation of galactic cosmic rays into the inner heliosphere. During 1978 several large solar wind blasts were observed at the two Pioneers. Shock waves associated with solar streams produced a large decrease in the cosmic ray intensity, beginning at about the Earth's orbit and reaching to about 16 times Earth's distance from the Sun, where an abrupt decrease in intensity was observed for the galactic cosmic rays and helium coming into the Solar System.

There were also strong 27-day variations in the intensities of galactic cosmic rays. Such variations were even more pronounced in measurements of the galactic helium, and they persisted for many months. These appeared to be associated with recurrent high-speed streams of the solar wind.

Material from the Stars

Uncharged hydrogen atoms from between the stars come into the Solar System along the plane of the Earth's orbit. There is no valid explanation yet as to why this occurs at 60 degrees from the direction of travel of the Solar System. Helium atoms discovered by the Pioneers are also believed to originate in interstellar space.

Changes in the ratios of helium, oxygen, and nitrogen to carbon among the low-energy cosmic ray particles were measured. They are believed to be caused by differences in the way the different elements are ionized by ultraviolet radiation from the Sun. When the interstellar atoms enter the heliosphere they are neutral, i.e. uncharged. Encountering ultraviolet radiation from the Sun they lose electrons at different rates in the process of becoming ionized. Helium, oxygen, and nitrogen penetrate most deeply before becoming ionized, and when each atom loses an electron, the atom is then accelerated by the turbulent magnetic fields carried by the solar wind and becomes a low-energy cosmic ray.

Experimenters on the Pioneer program hope that the spacecraft will still be operating and sending data to Earth when they reach some 30 to 40 times Earth's distance from the Sun in the late 1980s. Pioneer 11 travels in the direction the Solar System is traveling with respect to the local stars, toward a possible bow shock where the heliosphere—space surrounding the Sun dominated by the solar wind—holds off the interstellar medium, where interstellar particles and fields have motion relative to the Solar System. Pioneer 10 heads in almost the opposite direction.

In summary, Pioneer performed many experiments in interplanetary space between Earth and Jupiter and beyond. It mapped the magnetic field in interplanetary space and determined how the solar wind changes with distance from the Sun. Cosmic rays originating both from within and from outside the Solar System were observed and the spacecraft obtained information about interactions among these particles, the interplanetary field, and the solar wind.

Pioneer measured the amount of neutral hydrogen—non-ionized hydrogen atoms—in interplanetary space and in the vicinity of Jupiter. It ascertained how dust particles are distributed in the interplanetary space of the outer Solar System and determined the sizes, masses, fluxes, and velocities of small particles in the asteroid belt. It showed that the probability of damage by such particles to spacecraft passing through this region is very slight.

3.

Jupiter Flybys

MEANWHILE, Pioneer 10 approached its encounter with Jupiter. Its closest approach was scheduled for December 3, 1973, at 81,000 miles above the Jovian cloud tops. On November 6, at a distance of 15.5 million miles, the first pictures of Jupiter were obtained. Pioneer 10 crossed the orbit of Jupiter's outermost known satellite, Hades, on November 8, as controllers started the long process of sending some 16,000 commands to the spacecraft for the 60-day encounter.

Robert R. Nunamaker, who had been operations manager for Pioneer, emphasized that the spacecraft "does not do anything of itself. It is flown by man. We receive data from it and determine the sequence of commands needed to control it. To do this we have to know the spacecraft very well and have to have foolproof procedures. We have to know the science instruments, too, and we must know how they are working. Two or three times each day an emer-

gency appears in the data and nine times out of ten it is here on the Earth."

When each Pioneer reached the Jovian system, quick action again became the order of the day for these mission controllers. But now it was different from that during the launch phase because the spacecraft was over 500 million miles from Earth and radio signals took 92 minutes for the round trip. All command actions had to be planned well in advance because of this communications delay.

About noon on November 26, at a distance of about 4 million miles, farther from Jupiter than had been expected, Pioneer 10 passed through the bow shock where Jupiter's magnetic field deflects the solar wind. Before Pioneer encountered the shock, instruments on the spacecraft measuring the solar wind showed that it was blowing through space at 280 miles per second. After the spacecraft passed through the shock the velocity of the solar wind had dropped to only 140 miles per second, but its temperature rose from about 50,000 to 500,000 Kelvins (a Kelvin is one degree on the absolute scale of temperature). There was, of course, no danger to the spacecraft from this high temperature because the highly rarefied plasma could not transfer significant quantities of heat to Pioneer.

One day later, and also at noon, Pioneer crossed the boundary between the bow shock and the magnetic field of Jupiter, called the magnetopause. As expected, the solar wind did not penetrate any farther toward Jupiter. But Pioneer discovered that the magnetopause of Jupiter is relatively close to the bow shock compared with Earth's magnetopause and Earth's bow shock.

This environment of Jupiter was, however, quite different from that of our planet in several ways. Near the boundary of Earth's magnetic field, all the strength that holds off the solar wind is due to the Earth's field. But for Jupiter much plasma was found to be contained within Jupiter's magnetic field near the boundary, where it helps to hold off the solar wind. This additional barrier at Jupiter was found to be about equal to the magnetic field itself.

November 29, 1973: all spacecraft systems were operating perfectly; Pioneer crossed the orbits of all seven outermost known satellites of the Jovian system; soon afterwards it would plunge toward the radiation belts and its close encounter with the giant. Optimism was still high, for even if the Jupiter passage should damage some equipment, there was backup equipment available for the important post-encounter period of exploratory flight beyond Jupiter. This backup equipment had not had to be used on the long flight from Earth.

Most critical from the standpoint of radiation interfering with its operation was the photopolarimeter used to obtain the images of Jupiter. This instrument required long sequences of commands during the encounter to make the best use of time as the spacecraft hurtled by the giant. If the radiation should

generate false commands, the whole imaging sequence would have been upset and many important pictures of Jupiter would have been missed. Richard O. Fimmel, science chief for the mission to Jupiter, designed a sequence of contingency commands which would reconfigure the Pioneer spacecraft and its instruments on a continuing basis during the encounter. If a spurious command should be generated by the build-up of electrical charges or by intense radiation during the close approach to Jupiter it would be countermanded by a contingency command, so that the photopolarimeter would return to its preplanned imaging sequence.

The excitement of the encounter grew rapidly among the many people gathered at Ames Research Center, where scientists and science reporters waited to witness this first encounter with the giant of our Solar System.

Dr. Hans Mark, then Director of Ames Research Center, told newsmen: "This is an unusual event. The planet Jupiter. . . . is an object that has been the subject of fairly extensive observation for almost 400 years. Galileo, who looked at the planet through his primitive telescope in 1610, discovered . . . the four brilliant moons that surround the planet. This observation provided, I think, the first really visible proof, . . . that the Copernican model of the Solar System wasn't exactly the way it looks. Jupiter, therefore, served, perhaps, the function of quite profoundly changing the way we think about the universe."

Pictures of Jupiter were by this time coming back to Earth from the spacecraft and showing intriguing details of the Jovian cloud systems (figure 3-1). These pictures were displayed on television screens in the auditorium at NASA Ames by an image converter system which provided a quick display of the images, but they were not yet corrected for the motion of the spacecraft and the rotation of the planet. The system, developed by L. Ralph Baker of the University of Arizona, presented the images of Jupiter very quickly after they were received on Earth so that scientists could monitor operation of the imaging

Figure 3-1: By November 1973 Pioneer images of Jupiter began to reveal details of the cloud systems. The left image is in blue light and shows the Great Red Spot clearly. On the right image, taken in red light, the spot is hardly visible. These pictures were obtained at a range of 2,830,000 miles.

(NASA)

photopolarimeter during encounter. The system's video signal also was supplied to television networks so that many people were able to view the results of this first exploration of the huge planet on their home television screens as it took place half a billion miles from Earth.

The red and blue images received from the spacecraft provided good scientific data but if combined they did not produce a visually satisfactory image because they resulted in a purplish picture of Jupiter. So they were mixed as monochrome images to make a synthetic green signal. Then the red, blue, and synthetic green images were used to generate a normal three-color picture for display on monitors and for producing colored prints.

Initially the images of Jupiter were similar to those obtained by telescope from Earth. By December 2, 1973, however, the images exceeded the best Earth-based pictures (figure 3-2). When Pioneer 10 reached a distance from Jupiter of six times the planet's radius and was still operating without fault, it cleared the way for the 1977 Voyager mission, which was planned for a closest approach of six Jupiter radii. As detailed in a later chapter, one of these spacecraft did in fact approach closer than six radii from Jupiter.

But Pioneer 10 penetrated even farther than Voyager into the hostile environment of the Solar System giant; to two Jupiter radii above the cloud tops. Robert Kraemer, head of planetary programs at NASA Headquarters, commented: "We sent Pioneer 10 off to tweak a dragon's tail, and it did that and more. It gave it a really good yank and . . . managed to survive."

Said Charles Hall: "I always thought that the experimenters were exaggerating the story about radiation. I really couldn't believe it myself. But about six hours

Figure 3-2: Three images obtained in early December 1973, showing increasing detail as the range narrowed from 1,600,000 (left) to 1,140,000 miles (right).   (NASA/Ames)

before Pioneer 10 went through the belts I went into Room 209 [at Ames Research Center] and I walked into what looked like a funeral parlor." A group of experimenters were sitting around a table looking with gloom at data showing the levels of radiation. One of them exclaimed somberly, "I think we've had it." Hall relates how the experimenter showed him a curve which was extrapolated to show the radiation intensity now expected at the closest approach of Pioneer. It would be disastrous if true. The radiation levels at Jupiter were turning out to be much greater than had been anticipated by the scientists before the mission. Hall said he was really shocked. "If the radiation continued to increase [to periapsis] we were in serious trouble. It was too late to do anything." The spacecraft sped along its preplanned trajectory as the scientists and project personnel waited anxiously. "But within an hour," said Hall, "the radiation started falling off. We reached a peak long before we got to the closest approach. That was the most frightening time of the whole mission."

Until 10:00 A.M. on December 3 the imaging photopolarimeter functioned normally. Then, at a distance of nine Jupiter radii, the instrument started to act as though receiving commands to change the imaging sequence. The problem was quickly overcome by commands already placed in the memory of the spacecraft for this purpose, and good images of the terminator—the boundary between night and day on the planet—and of the Great Red Spot were obtained (figure 3-3).

Pioneer 10 then passed behind Jupiter and communication with Earth ceased. Now there was an anxious period of waiting for Pioneer to emerge from behind Jupiter, heralded by its radio signals again being received at Earth. The question in everyone's minds was whether or not all the scientific instruments would continue to work after their exposure to the intense radiation during periapsis passage.

In the press room everyone watched the display screens, waiting for a sign of a successful emergence of the spacecraft from behind the giant planet. A bright spot appeared, followed rapidly by others until a whole line was visible across the screen. The

a

Figure 3-3: Pioneer would fly past the Great Red Spot. Unfortunately it was traveling much too quickly to get good pictures when at its closest to the planet as shown in the artist's concept (a).

(b) Details begin to be revealed in the Great Spot as Pioneer approaches Jupiter.

(c) One of the better images obtained by Pioneer 10 of the spot shows swirling markings faintly within it. (NASA/Ames)

b

c

image when completed showed a thin crescent. The spacecraft was seeing sunrise on Jupiter. Pioneer 10 had survived through periapsis.

Some of the spacecraft systems showed the effect of the intense radiation, but although there were those anxious moments as the radiation reached levels that prevented some of the instruments from returning data, the instruments recovered and sent data again as soon as the high intensity dropped. A special Jovian radiation belt detector, designed in anticipation of inordinately high levels of radiation, was able to measure protons for the first time in the radiation environment of the giant planet. And to measure electrons, the experimenters had included a pair of special detectors.

In the hours following, more of these unique crescent views of Jupiter were obtained as Pioneer 10 headed away from the giant planet (figure 3-4). Richard O. Fimmel, the project's science chief who had been responsible for organizing all the commands to the scientific instruments aboard Pioneer and seeing to it that they were carried out by the spacecraft, commented just after the encounter: "This has been the most exciting day of my life!" The infrared radiometer had problems with many false commands generated by the radiation. Fimmel said: "I set people to work to pump out about 80 or 90 commands to that instrument in a hurry, and commands to fix up other instruments too."

But despite the radiation Pioneer 10 had, indeed, achieved its scientific objectives. It had shown that

a

b

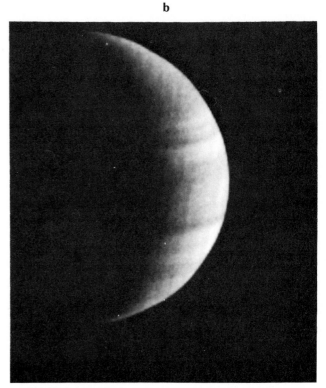

Figure 3-4: After flying close to the giant planet, Pioneer 10 then hurtled on toward the boundaries of the Solar System. As it did so it witnessed sunrise on Jupiter as the spacecraft emerged from the shadow of the planet.

(a) Artist's concept of Pioneer 10 viewing Jupiter as the planet can never be seen from Earth.   (NASA/Ames)
(b) Image of crescent Jupiter returned from the spacecraft.   (NASA)

NASA ARC JUPITER UNIV. OF ARIZONA
PIONEER 10 DATE: 12/5/73 TIME: 10:02:02
BLUE IMAGE RANGE:28.0RJ PHASE ANGLE: 111.6 DEG
PIONEER 10 IMAGE B26 BLUE

spacecraft could explore Jupiter and survive the hazards of the Jovian environment. We had now discovered what that environment really is and we had enough new data to whet our appetites for more exploration of the giant planets of our Solar System.

After its encounter with Jupiter, Pioneer 10 headed for the outer reaches of the Solar System, crossing Saturn's orbit in 1976 and the orbit of Uranus in 1980. The spacecraft would then travel at 25,000 mph into interstellar space, Until communications are lost far out in space it continues to gather information about the behavior of the solar wind at extreme distances from the Sun, measures integrated starlight from the Galaxy, and records cosmic rays.

Meanwhile, Pioneer 11 headed toward its encounter with Jupiter. On April 19, 1974, Pioneer 11's thrusters added 210.2 feet per second to the spacecraft's velocity, to aim the spacecraft so that it would penetrate deeper into the radiation belts and skim by the planet only 26,725 miles above Jupiter's turbulent cloud tops.

The inner radiation belt of the planet could seriously damage the spacecraft's electronics if the intensity of particles continued to increase beyond the maximum measured by Pioneer 10 at its closest approach of 82,000 miles above the cloud tops. The billions of electrons and protons trapped in Jupiter's magnetic field are concentrated toward the equatorial plane. Pioneer 10 traveled along the magnetic equator, where most energetic particles are concentrated. So its total dosage was high. Pioneer 11 was aimed to approach the radiation slowly, then pass through the maximum radiation quickly.

Pioneer 11's path had been chosen so that the spacecraft would approach Jupiter from below the planet's south pole. As it flew by the giant planet, the spacecraft would hurtle almost straight up so as to cross the intense radiation belts quickly and thereby reduce the time of exposure to the radiation and therefore the total dosage.

The close approach of Pioneer 11 would accelerate the spacecraft to a velocity of 55 times that of the muzzle velocity of a high-speed rifle bullet—to 108,000 mph, and this high speed would hurtle the spacecraft 1.5 billion miles across the Solar System to Saturn.

The path by Jupiter chosen to minimize radiation dosage also allowed the experimenters to obtain images of higher latitudes, including the polar regions, and to scan them with the infrared radiometer. Such views of Jupiter cannot be obtained from Earth. Scientists had stressed the scientific value of polar views of Jupiter at the first science meeting in 1970. Now Pioneer 11 would obtain such views. The south polar regions would be presented to the spacecraft prior to closest approach, and the north polar regions afterward (figure 3-5).

For several hours around closest approach of Pioneer 10 the viewpoint of Jupiter was nearly unchanging. This was because the spacecraft's direction of travel was the same as the direction of rotation of Jupiter. By contrast, Pioneer 11 traveled oppositely to Jupiter's rotation so that the spacecraft traversed a full circle of longitude of the planet during its close observations in the four hours centered on the time of periapsis. Pioneer 11 first passed in front of the planet, then around the dark hemisphere, and completed a circuit by crossing the incoming trajectory before heading out from Jupiter toward Saturn. Before closest approach to Jupiter, the view of the planet from the spacecraft showed the terminator near the left-hand edge of the disk. When speeding away from the planet, the spacecraft's view showed the terminator very near the upper right-hand edge of the disk. The trajectory chosen for Pioneer 11 also allowed the spacecraft to explore the magnetic field and radiation environment to higher latitudes of the magnetosphere.

The encounter of Pioneer 11 with Jupiter started November 3, 1974, when the spacecraft passed from interplanetary space into the Jovian system at a distance of about 15 million miles from the planet. Pioneer crossed the orbit of Hades, the outermost satellite of Jupiter, on November 6, but even at the enormous speed of the spacecraft it would not cross the orbits of the Galilean satellites until December 1, the day before closest approach. On November 25,

a

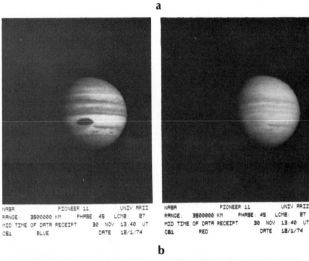

b

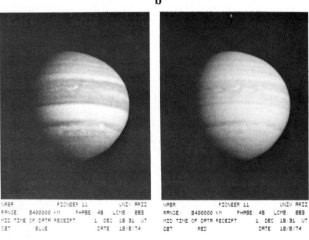

Figure 3-5: Images of Jupiter from Pioneer 11; top at 2,230,000 miles, and bottom at 1,500,000 miles.

(a)   This pair of images, blue on left and red on right, shows the Great Red Spot almost one year after the Pioneer 10 images, and some changes are apparent.

(b)   This image pair, blue on left and red on right, shows the other hemisphere of Jupiter with its equatorial plume. This too, has changed in the year since the Pioneer 10 encounter.
(NASA/Ames)

1974, Pioneer 11 crossed the bow shock and entered the inner system. Almost one day later, at a distance of 4.3 million miles from Jupiter, the spacecraft entered the magnetosphere where the magnetic field of the planet stops the solar wind from approaching any closer. But a short while later Pioneer 11 moved out of the magnetosphere for five-and-a-half hours

before crossing the bow shock once more and returning into the magnetosphere at 4 million miles from Jupiter. Then followed repeated bow shock crossings which, like those experienced by Pioneer 10, confirmed the model of the Jovian magnetosphere that likened it to an unstable soft balloon buffeted by the solar wind and often squeezed toward Jupiter on the side facing the Sun.

Many commands transmitted to Pioneer 11 on each of two days at closest approach were to ensure that the equipment carried by the spacecraft continued to operate correctly in spite of the effects of radiation. These commands repeatedly instructed the spacecraft to use the correct data format and the correct rate at which data were transmitted (data bit rate), and to keep its transmitter switched on and the scientific instruments operating. The spin-scan imaging photopolarimeter of Pioneer 10 lost several important images during encounter because the radiation generated false commands. To avoid this problem on the Pioneer 11 flyby, the instrument was periodically reset to the correct aspect angle for obtaining images of Jupiter.

But even more serious than the radiation environment of Jupiter, an industrial dispute in Austrialia imperiled the mission. Diesel operators supporting the Deep Space Tracking Station there threatened to strike. This would have put the important ground receiving station near Canberra out of action for the citical period of encounter. While negotiations proceeded, project management took precautions to save the 6 to 8 hours of encounter data vital to the science objectives of the mission. The encounter sequence was reprogrammed so that if necessary the Goldstone Tracking Station in California could maintain contact with the spacecraft during this vital period. Because the spacecraft would be very close to the horizon at Goldstone, the bit rate of data transmission to Earth had to be reduced from 2048 to 1024 bits per second if Canberra could not be used. Fortunately, negotiations were completed to allow technical personnel to operate the ground station.

The spacecraft went behind Jupiter at 9:01 P.M. PST on December 2, this was 21 minutes before its closest approach to the planet. The telemetered signals continued until 9:42 P.M. because of the time delay in transmission over the great distance to Earth. Pioneer 11 was unable to communicate with Earth at the critical time of closest approach. During this period of occultation, the spacecraft's memory recorded science data for later transmission to Earth. As with Pioneer 10 a year earlier, now came the anxious minutes when everyone had to wait . . . and hope. It was during this period that the spacecraft hurtled through its closest approach to the giant planet, skimming 26,725 miles above the cloud tops as it passed through the greatest intensity of the radiation belts.

At 9:44 P.M., 22 minutes after closest approach, Pioneer 11 was scheduled to emerge from behind Jupiter. But another 40 minutes passed while the radio signal traveled at 186,000 miles per second over the distance from Jupiter to the Earth. The signals, if they were still coming from the spacecraft, would not arrive at Earth until 10:24 P.M. Eleven seconds after 10:24 P.M. PST, the Deep Space Network station at Canberra, Australia, picked up the signal from the spacecraft and relayed it to the Pioneer Mission Operations Center at Ames Research Center. Then we knew here on Earth that Pioneer 11 had passed through its radiation bath. A cheer erupted spontaneously from the engineers, scientists, and newsmen covering the event. Another ten seconds elapsed, then the big antenna at Goldstone in the Mojave Desert picked up the signal too. Pioneer 11 had rushed in and successfully tugged at the tail of the giant dragon of the Solar System. But it had not emerged entirely unscathed by its encounter. The plasma analyzer, the infrared radiometer, the meteoroid detector, and the spin-scan imaging system had all delivered anomalous data. In addition, the output current from the power subsystem of the spacecraft fell slightly. Fortunately all instruments functioned correctly when the spacecraft settled down after leaving the intense radiation.

The most serious problem was an anomaly in operation of the infrared radiometer which caused it to miss observing the northern hemisphere of Jupiter. As soon as the signals reached Earth and the problem

was detected commands were sent to the spacecraft to reset the radiometer and thereby obtain about 50 percent of the initially planned observations of the northern hemisphere.

The intensity of high-energy electrons predicted for the close encounter was confirmed, but the proton flux was about ten times less than predicted. Near the planet radiation is intense but occupies a smaller volume than expected. The shells of extremely high-energy protons near Jupiter's magnetic equator are dangerous to spacecraft only at low latitudes. The intensity of high-energy electrons was only slightly higher than that found by Pioneer 10, even though Pioneer 11 went three times closer to the planet. However, Pioneer 11 discovered that the flux at higher latitudes is greater than expected from the measurements made by Pioneer 10.

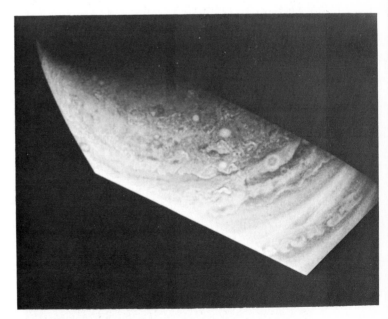

Figure 3-6: This Pioneer 11 picture covers part of Jupiter's north temperate zone and its north polar region. It shows the breakup of the regular banded structure of Jupiter's clouds at high northern latitudes. The polar regions contain what appear to be unorganized convective storms, many of which are circular and several hundred miles in diameter. The spacecraft was 26,000 miles from Jupiter when it obtained this image.
(NASA/Ames)

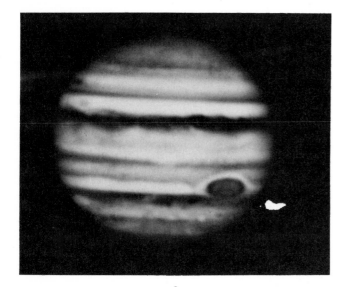

a

b

c

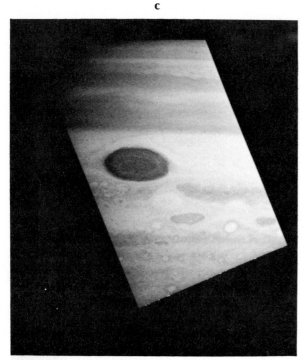

Figure 3-7: Changes in the Red Spot and its environs are shown in these three pictures.

(a)   Taken in 1969 from Earth.   (NASA)
(b)   Taken by Pioneer 10, December 1973.   (NASA/Ames)
(c)   Taken by Pioneer 11, December 1974. In this picture there is a bright oval and a dark elongated feature below the Great Red Spot, neither of which appear on the image obtained a year earlier.   (NASA/Ames)

Pioneer 11's instruments discovered that Jupiter's cloud tops are lower at the poles than at the equator, and are covered by a thick transparent atmosphere. There is much less evidence of rapid atmospheric circulation at the poles than at the equator, but there are many small convective cells which were not expected (figure 3-6).

There were many more flow features observed in the clouds around the Great Red Spot than were seen a year previously during the Pioneer 10 flyby (figure 3-7). New details within the spot seemed to be patterns of convection and circulation.

Immediately after the encounter with Jupiter, Pioneer 11 was renamed Pioneer Saturn. Its survival of encounter with Jupiter and the trajectory it followed allowed the spacecraft to head for an exploration of the next outer giant of our Solar System. Pioneer would be the first spacecraft to explore the strange planet and its stupendous ring system.

System Within a System

Radio observations of the paths of the two Pioneer spacecraft through the Jovian system were used to show that Jupiter and its satellites are heavier (by about half of the mass of Earth's Moon) than previously calculated from Earth-based observations of the Jovian system. Jupiter has a mass of 317 Earth masses and is about one-quarter of a Moon mass heavier than previously calculated. (The mass of the Moon is about 1/81 that of Earth.) Jupiter's gravity field reveals a symmetrical planet with no gravitational anomalies like those of Earth and the other inner planets. Jupiter thus appears to be an entirely liquid planet.

Jupiter's diameter and the amount of polar flattening were determined with greater precision by timing the occultations of the two spacecraft as each was occulted by Jupiter and then emerged from behind the planet. Jupiter is slightly more flattened at the poles than when measured from Earth. At a pressure of 800 millibars, which is near to the cloud tops, the equatorial diameter is 88,732 miles and the polar diameter is 82,980 miles. The polar flattening of Jupiter is thus twenty times as much as that of Earth. This results from Jupiter's fluid state, coupled with its high rotational speed. Average density of the planet, as calculated from its mass and volume, is 1.33 times that of water.

The Big Satellites

The Pioneers provided new measurements of mass and density for each of the Galilean satellites. Io is 1.21 lunar masses, Europa, 0.65, Ganymede, 2.02, and Callisto 1.46. Pioneer's measurement of the mass

of Io is 23 percent greater than that calculated beforehand. The density of the satellites was refined as the result of the Pioneer observations. Io's density is 3.52 times that of water; Europa's, 3.28; Ganymede's, 1.95; and Callisto's, 1.63. Thus the densities decrease with increasing distance from Jupiter. The two inner satellites appear to be rocky worlds. Indeed, the density of Io is greater than that of Earth's Moon. If the outer satellites are mainly water ice this would account for their low density. All the Galilean satellites are very cold. The average of their daylight surface temperatures is a frigid 230 degrees below zero Fahrenheit.

These new measurements threw light on how the Jovian system formed. The evidence points to the satellites having formed in such a way that lighter elements were depleted in the satellites close to Jupiter. However, it could be that although there was water in the inner part of the Jovian system, it did not condense on Io and Europa. They could have been too hot because of their closeness to glowing Jupiter during the period when the original nebula condensed, or later when Jupiter became very hot as it contracted gravitationally.

The two Pioneers obtained the first pictures from space of the Galilean satellites. The images were not clear (figure 3-8) but they hinted at a wealth of detail on these intriguing worlds. We had to wait another five years for the spectacular pictures from NASA's Voyager spacecraft, which confirmed the presence of large dark circular features, craters, valleys, and active volcanoes, and showed that there are widely different surfaces on each of the four worlds.

Emissions of sodium vapor from Io were detected by observations from Earth. They are from a cloud of sodium vapor that extends 10,000 miles from Io's surface. Pioneer 10's encounter with Jupiter was timed so that the spacecraft would be occulted by Io. As a result, the spacecraft's radio signals were able to probe close to the satellite. An ionosphere was discovered extending about 420 miles above the dayside of Io. The remarkable thing about this discovery is that Io possesses an ionosphere while immersed in the magnetic field of Jupiter. The density of this

Figure 3-8: The Galilean satellites as seen by the Pioneers. Although the spacecraft did not pass close enough to any of the satellites to reveal details of their surfaces, some images were obtained which suggested tantalizing features.

(a)   Io from a distance of 470,000 miles, Pioneer 11.

(b)   Europa from a distance of 220,130 miles, Pioneer 10.

(c)   Ganymede from a distance of 459,000 miles, Pioneer 11.

(d)   Callisto from a distance of 489,000 miles, Pioneer 11.

(NASA/Ames)

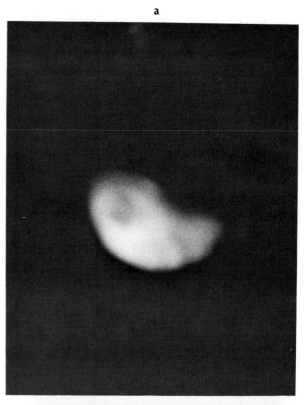

a

ionosphere varies from 60,000 electrons per cubic centimeter on the day side, to 9,000 on the night side of the satellite. Although the radio experiment did not detect an atmosphere on Io, the presence of an ionosphere implies that Io must have an atmosphere but one which is extremely tenuous. This atmosphere has a surface pressure of about one-hundred millionth Earth's sea-level pressure. Io's atmosphere probably consists of sulfur dioxide gas released from the volcanoes on the satellite which were discovered later by Voyager.

A surprising discovery by Pioneer 10 was that Io is embedded in a cloud of hydrogen which extends a third of the way around its orbit. The cloud is about 100,000 miles wide and high. It is 500,000 miles long and shaped like one-third of a doughnut. The cloud moves around Jupiter with Io at a distance of 250,000 miles from Jupiter. Pioneer 11 looked for hydrogen clouds around the other satellites but did not find any.

Particles and Fields

Jupiter's bow shock is produced when the high speed solar wind, carrying a magnetic field, interacts with the magnetic field of the planet (figure 3-9), and is abruptly slowed. Jupiter's bow shock is typically over 16 million miles across in the plane of the ecliptic. A magnetosphere surrounding Jupiter is where the magnetic field of the planet is dominant. This magnetosphere protects the planet from the solar wind. Pioneer 11 discovered that the magnetosphere is blunt on the sunward side and extends at least 80

b

c

d

radii of Jupiter above and below the planet. Between the magnetosphere and the bow shock is a turbulent magnetosheath where the solar wind is deflected around the magnetosphere.

Jupiter's magnetosphere rotates with the planet at several hundred thousand miles per hour. It has distinct regions. An inner region is shaped like a doughnut with Jupiter in the hole. Outside is a ringlike region caused by ionized gas thrown out into space by the rapid rotation of the magnetosphere. An outermost region is extremely unstable and pulsates under changing pressures of the solar wind, sometimes being pushed inward to half its size. On Pioneer 10's way into the Jovian system the spacecraft crossed the bow shock many times, and again on its outbound trajectory. Pioneer 11 reported three crossings inbound and three outbound, showing a very dynamic interaction between the solar wind and the magnetosphere of the giant planet.

The three regions of the Jovian magnetosphere revealed by the Pioneers are identified in figure 3-10. Under average conditions, the inner magnetosphere lies within about 20 radii of Jupiter, where the magnetic field of the planet predominates. In this inner magnetosphere the planetary magnetic field urges particles to move symmetrically about the magnetic equator. The middle magnetosphere extends from 20 to 60 Jupiter radii, where the magnetic field of the planet is severely distorted by trapped energetic particles moving at high speeds. In this middle magnetosphere, ionized particles form an electric current sheet around Jupiter. Here the particles move parallel to the equatorial plane of the planet. The current flowing in this sheet produces a magnetic field which at large distances from Jupiter is stronger than the magnetic field of the planet. The outer magnetosphere is beyond 60 Jupiter radii. This region is characterized by great irregularities in the magnitude and direction of the planetary magnetic field.

As Pioneer 10 hurtled toward its encounter with Jupiter, bursts of low-energy electrons were observed escaping from the giant planet. Such bursts of electrons had actually been seen at the orbit of Mercury

JUPITER'S MAGNETOSPHERE

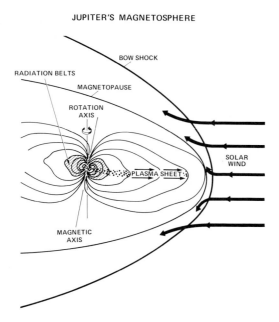

Figure 3-9: This diagram shows the relationship of Jupiter's magnetosphere to the solar wind. Between the bow shock and the magnetopause the solar wind is deflected around the planet by Jupiter's magnetic field. When the solar wind is more intense it pushes the bow shock in toward the planet.

(NASA/Ames)

but had not been recognized as originating from Jupiter. Low-energy electron fluxes, called "quiet time electron events" have also been observed at Earth for many years. They are now explained as being of Jovian origin. In fact, electrons from Jupiter are the most common interplanetary electrons in the inner Solar System except when there are large solar flares.

The Pioneers showed that Jupiter's magnetic field at the cloud tops is over 10 times as strong as Earth's field, with the total energy of the field some 400 million times that in Earth's field. (The magnetic moment is 20,000 times Earth's.)

Jupiter's field is tilted nearly 11 degrees to the planet's axis of rotation. The tilt of the field causes the radiation belts of Jupiter to wobble up and down in the surrounding space as the planet rotates. The result is that any spacecraft traveling passed the planet moves in and out of the radiation belts. Also

the center of the field is offset from the spin axis so that the strength of the field emerging from the cloud tops of Jupiter varies from place to place, 14 and 11 gauss in north and south polar regions respectively (Earth's polar field is 0.5 gauss). The poles of Jupiter's field are reversed compared with those of the Earth.

The field is distorted by the ring current in the magnetosphere and also by the field carried by the solar wind. Experimenters theorized that trapped particles forming the radiation belts are affected by the distorted field to produce a periodic release of relativistic electrons from Jupiter into interplanetary space.

The planet's intrinsic magnetic field is more complex than that of Earth. Within about three Jupiter radii the magnetic field seems more complex than a simple dipole field. Pioneer's magnetometer measurements fit a model in which more complex moments are at least 20 percent of the dipole moment, compared with only about 11 percent for the Earth. These larger moments tell us something about the interior of the planet. If the field is generated by an internal

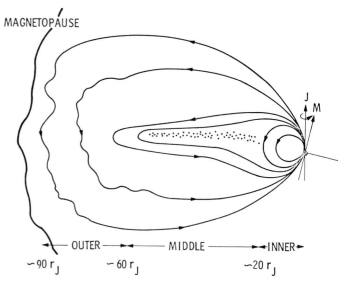

Figure 3-10: The three regions of the Jovian magnetosphere are identified; an inner region extends to about 20 radii of Jupiter, a middle region to about 60 radii, and the outer region extends to the magnetopause. All regions have quite different characteristics, as explained in the text.   (NASA/Ames)

dynamo, which is the generally accepted theory today, the divergence from a simple dipole field indicate complex circulation within the metallic hydrogen that forms a large part of the planet's interior. The high field strength also suggests that it is in the metallic hydrogen and not in a small rocky core that the field is generated.

As in the magnetosphere of Earth, energetic particles are trapped in the magnetosphere of Jupiter and produce radiation belts (figure 3-11). Scientists knew of these belts before Pioneer reached Jupiter from observations of decimetric radio waves generated by electrons in the belts. Indeed, their presence led scientists to predict that Pioneer would find that Jupiter has a magnetic field.

The inner, doughnut-shaped region of the magnetosphere, extending to about 10 radii of Jupiter, has not only electrons but also protons with a wide spectrum of energies. The intensity measured there by Pioneer was the highest natural radiation intensity ever measured and it was comparable to radiation from a nuclear bomb explosion in Earth's upper atmosphere. In the inner Jovian belt peak electron intensity is 10,000 times greater than that in Earth's belts, and protons are several thousand times as intense as in Earth's belts. The science experiments aboard Pioneer showed that particles in this inner region of the Jovian radiation belts most probably

originate from Jupiter and not from the solar wind, as do the particles in Earth's radiation belts.

Particles and fields measurements made by the Pioneers lead to a model in which Jupiter's radiation belts in the middle and outer magnetosphere are represented by a disk. Because the magnetic field lines near the magnetic equator are closed they are able to trap charged particles at distance of up to 100 times the radius of Jupiter. The outer radiation belt reaches to at least 100 radii of Jupiter. Its most energetic particles are high-energy electrons concentrated into a relatively flat area. Energetic particles revolve around Jupiter at the same rate as the planet. This co-rotation of energetic particles with Jupiter extends to the magnetopause, quite different from the Earth where co-rotation ends at the outer boundary of the plasmasphere, which is well within the magnetosphere. Also, the presence in the Jovian outer magnetosphere of a large number of electrons with energies greater than 20 MeV cannot be explained by a trapping of particles from the solar wind in this region either. If the particles did originate from the solar wind they could only possess energies of up to 1 keV. Capture of solar wind particles seems a relatively minor process within the Jovian magnetosphere. The dominant process seems to be acceleration of particles internally available within the magnetosphere—quite different from conditions in Earth's radiation belts. The solar wind is, however, necessary to generate conditions to transfer rotational energy from Jupiter to the charged particles. It is the flow of the solar wind past Jupiter that generates the axially asymmetric and non-rotating conditions essential to developing the magnetosphere of the giant planet.

An important discovery was that both electrons and protons stream from Jupiter along high latitude field lines. The electrons have energies between 40 and 560 keV, and the protons have energies between 0.61 and 3.41 MeV. This effect may indicate that energetic particles recirculate within the magnetosphere, gaining energy cyclically, and that the particles are ejected from Jupiter's magnetosphere into interplanetary space from higher latitudes and not exclu-

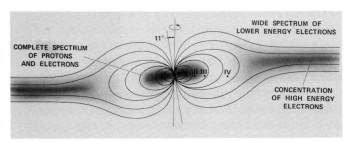

Figure 3-11: The locations of the Galilean satellites in the Jovian radiation belts lead to involved interactions between the satellites and the charged particles. While the satellites orbit Jupiter in the planet's equatorial plane, the magnetic axis is tilted so that the radiation belts wobble around the planet as it rotates. This too leads to involved interactions with the satellites.   (NASA/Ames)

drogen, with about 10 to 15 percent helium and traces of other gases. Because Jupiter was thought to be very similar in composition to the Sun, scientists believed that helium was present on Jupiter. But it remained for Pioneer to identify helium on Jupiter positively. Jupiter's atmosphere also contains small amounts of ammonia and methane, and traces of deuterium, acetylene, ethane, and phosphine, all of which have been detected from Earth. Water vapor is also present in very small quantities, and there is also carbon monoxide, hydrogen cyanide, and germane.

The Turbulent Atmosphere: Weather and Clouds

Like clouds on our own planet, Jupiter's clouds form in the atmosphere by condensation. But they are of ammonia and ammonia compounds as well as water. The topmost clouds are probably ammonia crystals. Water clouds are lower in the atmosphere. An inversion layer is 22 miles above the visible clouds. This layer results from the action of a layer of aerosols and hydrocarbons such as ethane and acetylene which absorbs incoming solar radiation. Pioneer's occultation experiments reveal the temperature inversion at a level in the Jovian atmosphere where the pressure is between 10 and 100 millibars. The temperature is between −207° and −171° F at 10 millibars, and −297° to −261° F at 100 millibars. At the cloud tops the temperature is about −234° F.

Because Jupiter's axis is tilted only slightly, the planet does not have seasons like the Earth and Saturn. Solar radiation per unit area of Jupiter is more concentrated equatorially than at the poles. Unlike Earth and other terrestrial planets, the equator of Jupiter is not warmer than the poles. Infrared measurements by the Pioneers showed that, in fact, temperature is fairly constant between the poles and the equator, between north and south hemispheres, and between day and night sides. One theory to explain this even temperature distribution is based on the very efficient circulation within the atmosphere to redistribute solar heat. More likely, however, because no flow of atmosphere from equator to poles

is apparent in the cloud patterns of Jupiter, the explanation is that heat flux from within Jupiter is higher at the poles so that it balances the lack of solar input there. Convection over the planet eliminates any temperature differences. At the poles convection brings more heat from the interior of the planet to compensate for the lower solar input in polar regions. At the equator convection is much reduced. Thus Jupiter behaves as though controlled by a natural thermostat.

The question of the cause of the Great Red Spot and why it has lasted so long was not answered by Pioneer, although there were some pointers to answers. Atmospheric flow on a rapidly rotating planet with an internal heat source has been simulated on powerful computers and some mathematical models develop features similar to the Great Red Spot. Time-lapse motion pictures developed from the imaging System of the Voyager spacecraft have provided more information about the complex motions within the spot, as described in a later chapter.

Pioneer 10 made a significant discovery by obtaining an image (figure 3-13) of a similar red spot in the northern hemisphere. Such spots had been seen from Earth but never in sufficient detail to show that they were similar in structure to the Great Red Spot. The small spot's shape and structure confirmed that red spots are hurricanelike features in the atmosphere.

The Pioneer images suggest that the Great Red Spot rotates counterclockwise; and this was later confirmed by the Voyager pictures. The spot is anticyclonic and seems to be an ascending mass of gas which flows outward several miles above the surrounding clouds. The color of the spot may be caused by phosphine carried to heights at which solar ultraviolet produces red phosphorus.

When the images of Jupiter were examined, there were surprises. Detailed cloud structures in intermediate latitudes were not expected. Billows and whirls near the edges of belts and zones showed rapid changes in wind direction and wind speeds. Latitudinal and longitudinal motion was evidenced by trends and slants in the North Tropical Zone. A

Figure 3-13: Pioneer obtained this image of a small red spot in the northern hemisphere of Jupiter which shows swirls within it and cloud patterns to east and west. Such spots had been seen previously from Earth, but not in sufficient detail to recognize that they have characteristics very similar to the Gread Red Spot. They do not, however, last so long.

(NASA/Ames)

plume in the Equatorial Zone (figure 3-14) provided structural information that aids our understanding of this common cloud form there. Unique views of the south polar regions (figure 3-15) revealed oval and circular cellular structures, but the dark belts and light zones characteristic of regions closer to the equator are absent. Also the atmosphere appears thicker over the polar clouds than over temperate and equatorial regions.

Optical depth of the atmosphere above the cloud tops was three times as great at latitudes above 60 degrees than in the equatorial zone, possibly because of a thin high cloud layer or some as yet unknown upper atmosphere absorber.

Earlier theoretical work which suggested that the Great Red Spot and the light-colored zones are cloudy regions of swirling anticyclones and rising air masses seemed to be confirmed by the Pioneer results. The darker belts are cyclonic, sinking masses of atmosphere (figure 3-16). Belts and zones scatter sunlight differently. It is speculated that the belts

Figure 3-14: Periodically, streaming plumes are seen in the equatorial regions of Jupiter. Pioneer obtained good images of one of these features. It has a bright nucleus, probably of rising masses of atmosphere. The streamer stretches from the nucleus about 40,000 miles.

(a) Top. Taken at a range of 800,000 miles.

(b) Bottom. Taken at a range of 451,000 miles. (NASA/Ames)

Figure 3-15: The trajectory of Pioneer 11 was chosen to fly rapidly through the radiation belts, south to north. This permitted views of the south polar regions on approach, and of the north polar regions on the outbound trajectory. This image was obtained when Pioneer 11 was 750,000 miles from Jupiter.   (NASA/Ames)

may appear dark because of dark aerosols suspended in the gaseous atmosphere there. On Jupiter cyclones and anticylcones are stretched into linear or hook-shaped features by the rapid rotation (figure 3-17). Extremely turbulent areas separate adjacent bands which are moving around the planet at different velocities.

Terrestrial storm systems persist for days or weeks, but storm systems on Jupiter last much longer. An extreme example is the Great Red Spot, which has persisted for centuries. The storm systems also move relative to one another. On Earth strong interactions between atmospheric systems and land masses dissipate energy and break up atmospheric systems.

Also, the terrestrial storm systems obtain their power from solar heating which is concentrated in the tropics. When a storm moves from the tropics it cannot replenish its energy. It fades away. But Jupiter's storms receive energy mainly from heat flowing from the interior of the planet, heat which is, moreover, evenly distributed between equator and poles and over day and night hemispheres. And there are no surface features like those on a solid planet to absorb energy from a storm system. These facts probably account for why the Jovian storms last so long.

Bright zones on Jupiter have been likened to tropical convergences on Earth—bands of thunderstorms, a few degrees north and south of the equator. The terrestrial features are caused by interactions between the trade winds, blowing toward the equator, and rising moist air in the tropics. The thunderstorms spread cirrus cloud tops, and air from them flows poleward. The rising air masses on Jupiter may produce masses of clouds which spread at the tops and produce the bright bands of the North Tropical and South Tropical Zones.

CIRCULATION OF BELTS AND ZONES

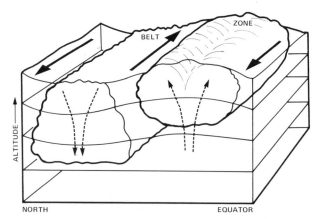

Figure 3-16: Evidence mounted during the Pioneer missions, and was supported by Earth-based infrared observations, that the zones are regions of warm, moist, rising atmosphere, and the belts are cool, dry, falling atmospheric masses. Coriolis forces urge the air masses into motion around the planet instead of from equator to pole as on Earth.   (NASA/Ames)

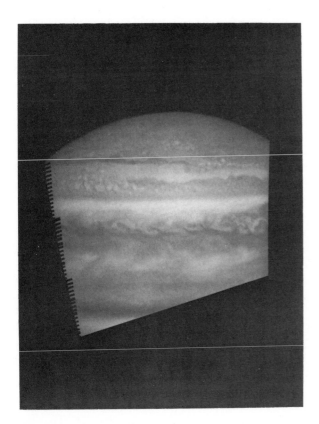

Figure 3-17: The images revealed hooklike features between belts and zones where masses of atmosphere are swirled into eddies by air masses moving in opposite directions relative to each other. (NASA/Ames)

But a major question is why the clouds are colored if they are condensations of ammonia and water. Both these substances are colorless when condensed. Some compounds of ammonia are colored when exposed to ultraviolet radiation. Sufficient solar ultraviolet radiation to do this penetrates to the cloud levels in the Jovian atmosphere. Possibly carbon compounds or traces of sulfur and phosphorus also contribute to the colors. Mere traces would react in sunlight and produce the colors such as those visible on Jupiter.

Colors can also be produced by the presence of free radicals. At the low temperatures in the high cloud layers, chemical compounds can be short of their

normal complement of atoms and still remain relatively stable. These compounds are called free radicals and they are generally highly colored.

The limb of the planet is darker than the rest of the disc. This indicates that above the dense lower layers of clouds there is a thin upper layer, semitransparent to red light. The particles of the upper clouds are much smaller than particles in Earth's clouds.

A detailed model of cloud layers of Jupiter was still being developed at the time of the Voyager encounters in 1979. There appear to be two main layers; a thick, low deck above which there is a gaseous atmosphere; and a thin, high deck topped by a layer or layers of aerosols. The Jovian cloud particles seem to be irregular in shape; most probably they are larger than the wavelength of light.

Ground-based infrared observations at wavelengths of five micrometers show belts and zones which are much the same as seen in visible light pictures of the planet. The dark visible belts are light (hotter) in the infrared pictures, and the light visible zones are dark (cooler). Infrared radiation from Jupiter originates deep within the atmosphere. Hence the dark visible belts must be lower, or thinner, hotter clouds, and the bright visible features must be higher, or thicker, cooler clouds. The Pioneer infrared maps of Jupiter were made at 20 and 40 micrometers. They confirm the high and low clouds and that Jupiter emits more heat than it receives from the Sun.

When radiation affected the infrared instrument some observations of the northern hemisphere were lost, but two good infrared spin-scan images of the planet were obtained. One, centered at 41 degrees south latitude, was a complete image. The other, centered at 52 degrees north latitude, was only a partial image. Jupiter does not appear to emit as much internal heat as was once thought. Both spacecraft made observations from which scientists derived the ratio of total thermal energy to absorbed solar energy as 1.7 plus or minus 0.2, compared with previous estimates of 2.5 plus or minus 0.5. The revised ratio supports the theory that Jupiter is losing internal energy by cooling and contraction, and not by gravitational separation of helium from hydrogen.

Overview

In summary, the Pioneers performed an involved series of experiments within the Jovian system and enlightened us about its true nature. The spacecraft mapped the magnetic field of the planet, determining its intensity, direction, and structure. Numbers of electrons and protons of various energies were measured along the trajectories of the spacecraft through the planetary magnetosphere to determine particle distribution, and the results showed how the particles are affected to the satellites. The spacecraft mapped the interactions of the planet with the solar wind.

The atmosphere of Jupiter was probed at occultation with S-band radio waves—as was Io, to investigate the characteristics of Io's tenuous atmosphere, in which ionospheric regions were identified. Auroras in the polar atmosphere of the planet were searched for, and information was obtained to help interpret the characteristics of the decimetric and decametric radio waves from Jupiter.

Temperatures of the atmosphere of Jupiter and of the surfaces of some of the large satellites were measured. The thermal structure of the planet was mapped by measuring infrared radiation. The amount of heat the planet radiates into space compared with the heat it receives from the Sun was determined more accurately.

Many spin-scan images of Jupiter were obtained in two colors during the encounter sequences, several being of special planetary features. Polarimetry measurements were made of Jupiter and some of the large satellites. Pioneer also investigated at relatively close range several of the Galilean satellites by spin-scan imaging and other measurements to determine their sizes and other physical characteristics.

The shape of the external gravitational field of Jupiter was determined, from which scientists inferred the internal mass distribution of Jupiter and the structure of its field. The spacecraft helped determine more precisely the masses of Jupiter and the Galilean satellites by providing accurate information about the effects of their gravitational fields on its motion. The flyby trajectory also provided information to calculate with greater precision the orbits and ephemerides of Jupiter and its Galilean satellites in preparation for the Voyager mission.

Most important, the Pioneers had described the system sufficiently for the more complex Voyager missions to seek answers to more specific questions and use high-resolution cameras to obtain images showing great detail within the planet's cloud systems and on the surfaces of the Galilean satellites.

4.

The Voyagers

THE name Voyager was applied to a planned space mission quite early in the Nation's space program. Originally, Voyager was a spacecraft system to explore Mars. It originated in 1965 when NASA requested industrial firms to participate in the design of an automated spacecraft to explore the planets starting with Mars. NASA emphasized at the time that although a scientific objective of Voyager was to land a life detection experiment on Mars, the spacecraft design was intended to support the physical exploration of the Solar System so that Voyager would be suitable for many different scientific experiments. Three companies were selected to design the spacecraft, and in July 1965 NASA assigned management of the landing capsule program to the Jet Propulsion Laboratory, Pasadena, California, where Dr. Donald P. Burcham was appointed project manager, and Harris M. (Bud) Schurmeier, formerly project manager for the Ranger unmanned lunar spacecraft, was appointed deputy project manager.

In July also, scientists were invited to participate by proposing experiments to be carried by Voyager. It was intended that the spacecraft should be launched by the huge Saturn V booster developed for the manned lunar landings. But by December 1965 the Voyager program had been deferred and smaller Mariner-type spacecraft were assigned to fly by Venus and Mars before any attempt would be made to send a spacecraft to land on Mars.

Voyager languished in limbo for several years, then in early 1967 it was resurrected when NASA organized a management team for long-term exploration of the planets using Voyager, a particular goal being to land on Mars and also survey that planet from orbit. Assignments were for Marshall Space Flight Center, Huntsville, Alabama, to be responsible for the orbiting vehicle, Langley Research Center, Hampton, Virginia, for developing a landing capsule, and the Jet Propulsion Laboratory for tracking and mission operations and for the surface laboratory. Said Edgar M. Cortright, Deputy Associate Administrator of NASA, in March 1967, "Voyager constitutes the most important undertaking in space exploration since Apollo was begun in 1961."

But the project fizzled. Congress decided to fund no major increases in the national planetary programs for 1967 and instead authorized more funds for space applications. As a result, NASA had to concentrate on ways to develop a less expensive program of planetary exploration without Voyager, using a Titan launch vehicle and the Mariner series of spacecraft. Ultimately a spacecraft named Viking made the unmanned landings on Mars in 1976.

But by 1970 scientists and space engineers were actively investigating ways to explore the outer planets during the decade. A unique planetary alignment which began in 1977 offered an opportunity to explore all the outer planets of the Solar System using single spacecraft (figure 4-1) to perform multiple flybys. This opportunity would not repeat for almost two centuries. The concept became known as the Grand Tour, which was mentioned in an earlier chapter in connection with the development of Pioneer. Studies showed the feasibility of a series of missions following the Pioneer missions. A Jupiter flyby and out-of-the-eclipitc probe could be launched June 23 to July 13, 1975. A Jupiter/Saturn/Pluto mission could be launched August 23 through September 12, 1977, and a Jupiter/Saturn/Uranus/Neptune Grand Tour could be launched in that same period. A Jupiter/Uranus/Neptune mission could be launched October 27 through November 16, 1969, and a Jupiter Orbiter might conclude the series with a launch between December 1 and 21, 1980. The Grand Tour was an ambitious and stimulating plan to extend Man's dominion to the borders of our Solar System. But the plan was not received well in the climate of dissatisfaction and distrust of technology and science festering as a result of the Vietnam war and the accompanying economic shortages which are the aftermath of all large military campaigns.

Nevertheless, in April 1971 NASA selected a team of 108 scientists from the United States and 6 foreign

Figure 4-1: A spacecraft for a Grand Tour mission sweeps by Jupiter in this artist's concept in early 1970 of the 11.5-year mission that would take it to the planets Jupiter, Saturn, Uranus, and Neptune. This concept was developed by North American Rockwell Space Division in studies of missions that would have taken advantage of the unique alignment of outer planets that began in 1976 and would not recur for about 180 years. (North American Rockwell)

countries to participate in defining the experiments for outer planets missions. That five hundred scientists had submitted proposals was evidence of great interest by the science community. Funds to start the Grand Tour missions were requested of Congress in the fiscal 1972 budget. But the Grand Tour did not receive congressional support partly because some scientists short-sightedly withheld their full support and contributed to encouraging popular rejection of high technology explorations. Also there was a legitimate doubt as to whether spacecraft could be designed reliably enough to function in space for a decade or more as required by the Grand Tour. NASA designed a reduced program for a Jupiter/Saturn mission which would require a spacecraft to function reliably for four years only. In February 1972, three years after giving its blessing to the Pioneer Jupiter program, NASA released details of the plan for a follow-on mission to the outer Solar System. Two Mariner-class spacecraft would be launched in 1977 on trajectories that would carry them by Jupiter and on to Saturn.

This became the Mariner Jupiter/Saturn project. The two-planet flyby replaced the more expensive Grand Tour that would have visited all five outer planets by two launches in 1977 and two more in 1979. The Grand Tour was finally dropped from the Nation's space program because of postwar budget constraints over the next several years. The Mariner Jupiter/Saturn spacecraft were to be launched by Titan/Centaur rockets. The spacecraft would take about one-and-a-half years to travel to Jupiter and about three-and-a-half years to reach Saturn. After encounter with Saturn they would escape from the Solar System. Project responsibility was assigned to the Jet Propulsion Laboratory where there had been, since 1968, a study team working on a concept for a thermoelectric outer planets spacecraft known by the acronym TOPS. But Mariner Jupiter/Saturn differed from TOPS, explains Schurmeier, because the radiation hardening was omitted from the Mariner design and the spacecraft would fly past Jupiter farther out than was planned for TOPS.

In July 1972, Dr. William H. Pickering, then director of the Jet Propulsion Laboratory, formed a project office at the laboratory for unmanned missions to Jupiter and Saturn. Harris Schurmeier was named manager of the new Mariner Jupiter/Saturn project, with E.J. Smith the acting project scientist and R.L. Heacock the spacecraft system manager. By the end of the year, NASA had appointed as project scientist Edward C. Stone, a distinguished magnetospheric physicist from California Institute of Technology. His job was to coordinate all the science activities of this tremendously complex mission.

Stone recalls: "I started on the program with the Grand Tour as an experimenter in 1970 when all the preplanning was going on. Mariner Jupiter/Saturn was a direct result of the Grand Tour studies. There was some concern about the cost of the Grand Tour, and out of this came the scaled-down program. It was more reasonable to build a spacecraft that would

operate for four years than one that would have to work for the ten years needed for the Grand Tour. . . . The science investigations were selected by NASA in the Fall of 1972. I was not involved in the selection process, probably because I was a co-investigator on one of the competing experiments. The way NASA selects experiments is to solicit proposals, then set up a number of peer group committees to evaluate the proposals. Out of that selection NASA then internally selects investigators."

In December 1972, NASA announced that 90 scientists in the United States and 4 foreign countries had been provisionally selected to participate in the mission. They were chosen from more than 200 scientists who had submitted proposals in response to NASA's invitation issued in April of 1972. These scientists were grouped into 11 investigative areas, each of which was to be represented on a Science Steering Group which was responsible for the overall science program and worked with mission planners to design the spacecraft and the details of the mission. Except for the imaging and radio science experiments, for which instrumentation was furnished by the project itself, individual science groups were each responsible for designing and constructing their own instruments.

Almost one year was spent making sure that the requirements for carrying each instrument on the spacecraft were fully understood. Then NASA confirmed the investigators in September 1973. At that time one experiment—to investigate micrometeorites in interplanetary space—was dropped because of an overall cost problem. An ultraviolet photometer was provisionally selected, said Stone, "then that experiment did not get confirmed because we had the Pioneer flyby of Jupiter which found that the radiation was a thousand times more intense then expected." To avoid radiation damage Mariner Jupiter/Saturn would have to fly too far from Jupiter to obtain useful science measurements. The whole spacecraft design had to be reevaluated. It was clear that Mariner Jupiter/Saturn's project management had to take a close look at all the instrumentation and the effects of the more intense radiation environment, and at the cost of making the experiments suitable for exposure to that environment by using shielding and hardened parts. The payload weight would be increased by such changes. About this time NASA Headquarters made the decision not to confirm the ultraviolet photometer but to include a plasma wave experiment which had been originally proposed but was not selected until then.

There were two launches planned, both for the period August-September 1977. The two spacecraft were identical and made maximum use of the subsystems designed for the Viking Orbiter, which was the most advanced spacecraft in the Mariner line to date.

The Mariner Jupiter/Saturn team of top-ranking space scientists, engineers, and mission planners had what seemed a short four years to convert the plans and technical papers into a pair of working spacecraft capable of traveling through space for many years and operating a battery of sophisticated scientific instruments across billions of miles of space. The mission offered a particular challenge to scientists and engineers to explore in depth the new regions of space opened by Pioneer. But also the mission offered a major challenge in designing the instruments to have long lifetimes and in handling the enormous quantities of data that would be received from them. NASA adopted a conservative approach by insisting on the use, wherever possible, of proven spacecraft hardware. This was done to increase the chance of success and to hold costs down.

There was important interplay between Pioneer and Mariner Jupiter/Saturn. Pioneer's tasks were to show that the asteroid belt could be penetrated safely, and to find out what the radiation environment of Jupiter was really like. Mariner Jupiter/Saturn was intended to get to Saturn, where there was the objective of finding out as much as possible about the rings. At Saturn there was the problem of where to pass through the ring plane, and Pioneer 11 offered the possibility of going first to Saturn to obtain data for this decision, which it did, as described later.

The new project anticipated learning much from the results of the Pioneer mission, which was at that time half-way to Jupiter. These results were expected to be available before the final design was firmed for the Mariner Jupiter/Saturn spacecraft. At this time science objectives were defined (see table 4-1), but were expected to be refined if necessary when the results from Pioneer became available. The Mariner

Table 4-1   Science Objectives of the Mariner Jupiter/Saturn Project When It Was Established in 1972

Imaging: Visually characterize the planets, their satellites, and the rings of Saturn.

Radio Science: Study the atmosphere of Jupiter and Saturn and the satellites, the size of particles in the rings of Saturn, and interplanetary physics, using the telecommunications systems of the spacecraft.

Infrared Spectrometry and Radiometry: Study both the global and local energy balance in conjunction with the measurement of reflected solar energy; determine atmospheric composition, including the ratio of hydrogen to helium and the abundance of methane and ammonia; and investigate the composition, thermal properties, and size of particles in Saturn's rings.

Ultraviolet spectroscopy: Analyze the atmospheres of Jupiter, Saturn, and encountered satellites for their major constituents, including the mixing ratio of hydrogen and helium and the thermal structure of each atmosphere.

Magnetometry: Study the magnetic fields of Jupiter and Saturn and the interplanetary magnetic field out to its boundary with the interstellar magnetic field, if possible.

Plasma: Determine the properties of the ions in the solar wind, obtaining accurate measurements of their velocity, density, and pressure, and determine how they interact with the planets.

Low-energy Charged Particles: Study the magnetosphere and trapped radiation belts of Jupiter and the possible magnetosphere and trapped radiation belts in the vicinity of Saturn; and separate and analyze the low-energy galactic cosmic ray particles.

Interstellar Cosmic Rays and Planetary Magnetospheres: Determine the energy and identity of galactic cosmic ray charged particles of medium to high energy, and of higher energy particles trapped in planetary fields.

Interplanetary and Interstellar Particulate Matter: Detect and determine the range and velocity of particulate matter by observing reflected solar radiation and counting particle impacts by their light flash and the sound of their impact (later cancelled).

Photopolarimetry: Measure reflection and scattering properties of particles in the atmospheres of Jupiter and Saturn, their satellites, and the rings of Saturn.

Planetary Radio Astronomy: Record and study planetary nonthermal emissions from Jupiter and Saturn, and plasma resonances in the magnetospheres of the planets.

philosophy was also to make the spacecraft adaptable so that experiment sequences could be changed after launch and even during flybys, to deal with targets of opportunity. If something interesting surfaced on the approach to Jupiter or Saturn the scientists would have the capability of reconfiguring their experiments to investigate the new phenomenon. This approach had paid dividends in all the earlier programs to explore Mars, and was expected to be even more valuable in exploring all the diverse planets and satellites of the outer Solar System with only two identical spacecraft.

In general, science objectives crystallized as follows: to conduct comparative studies of the Jupiter and Saturn systems, the environment, atmosphere, surface, and body characteristics of the planets and one or more of their satellites; to determine the nature of the rings of Saturn; and to study the interplanetary and interstellar media. To meet these objectives, science investigators were appointed as listed in table 4-2.

A principal investigator or team leader from each of the 11 investigations formed the Science Steering

Group which was chaired by Stone. They made recommendations on required trajectories, sequencing of experiments, and general science aspects of the project.

Each Mariner Jupiter/Saturn spacecraft would carry wide-angle and narrow-angle television cameras, cosmic ray detectors, infrared spectrometer and radiometer, low-energy charged particle detectors, magnetometers, a photopolarimeter, planetary radio astronomy instruments, plasma and plasma wave experiments, and an ultraviolet spectrometer.

The Jupiter/Saturn spacecraft was closely related to the Mariner spacecraft that had so successfully explored Mars. The stabilization and trajectory correction subsystems were applicable to the mission to the outer planets. Also, the computer and data handling equipment for acquiring, formatting, and storing the science and engineering data were similar. However, the specific requirements of the Jupiter/Saturn mission demanded some new design approaches. Most significant was the way in which electrical power for the spacecraft would be generated. As with the Pioneer spacecraft, practically sized arrays of solar cells were not adequate to meet power needs at great distances from the Sun. Radioisotope thermoelectric generators (RTGs) had to be used instead.

The Jet Propulsion Laboratory had not used RTGs on earlier spacecraft, but the power sources had been used on Pioneer and on DOD satellites. Those used for Mariner Jupiter/Saturn (figure 4-2) were a modified version of RTGs developed by the Department of Energy. The concern was that the generators were too heavy, so their weight had to be reduced. The other concern was degradation of output with time. Within the units temperatures reach 1800° F, and this high temperature difference across a thermocouple generates the electrical output. Anything that allows the heat to leak across the thermocouple reduces the electrical power output. Deterioration of the thermocouple material due to chemical instabilities accelerated by high temperature also reduces power output. The project engineers found that at the high temperatures experienced within the RTGs,

Table 4-2 Science Investigators for Mariner Jupiter/Saturn

Experiment	Investigator	Affiliation
Cosmic Ray	R.E. Vogt	California Institute of Technology
Imaging	B.A. Smith	University of Arizona
Infrared (IRIS)	R.A. Hanel	Goddard Space Flight Center
Low-energy Charged Particles	S.M. Krimigis	Johns Hopkins University
Magnetometry	N.F. Hess	Goddard Space Flight Center
Photopolarimetry	C.F. Lillie	University of Colorado
Radio Astronomy	J.W. Warwick	University of Colorado
Plasma Particles	H.S. Bridge	Massachusetts Institute of Technology
Plasma Waves	F.L. Scarf	TRW Systems Group
Radio Science	Von R. Eshleman	Stanford University
Ultraviolet Spectrometry	A.L. Broadfoot	Kitt Peak National Observatory

VOYAGER RADIOISOTOPE THERMOELECTRIC GENERATOR

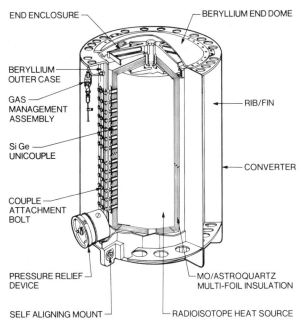

END ENCLOSURE

BERYLLIUM END DOME

BERYLLIUM OUTER CASE

GAS MANAGEMENT ASSEMBLY

RIB/FIN

Si Ge UNICOUPLE

CONVERTER

COUPLE ATTACHMENT BOLT

PRESSURE RELIEF DEVICE

MO/ASTROQUARTZ MULTI-FOIL INSULATION

SELF ALIGNING MOUNT

RADIOISOTOPE HEAT SOURCE

Figure 4-2: Advanced radioisotope thermoelectric generator developed for the Voyager spacecraft to provide electrical power at great distances from the Sun where solar cells are impractical. (JPL)

a silicon base material sublimed from the insulation. Later this material was deposited within the unit where it could produce a heat leak across the thermocouple. Moreover, the process deteriorated the thermocouple material too. The need was to reduce this effect so that it would not be too great over the duration of the expected mission to the outer planets. In a veritable race against time a nitride coating process was developed to repress the sublimation of the silicon base material. It worked fine. The output of the RTGs held up very well over the mission.

Instruments that had to make very sensitive measurements in interplanetary space and at the planets had to be located as far as possible from the main body of the spacecraft to minimize electromagnetic interference from the spacecraft and its subsystems.

As with the Pioneer spacecraft, these instruments were placed on long booms which were designed so that they could be unfurled from a stowed position within the spacecraft once it had separated from the launch vehicle. These magnetometer booms presented problems of placing the magnetometers accurately with respect to the spacecraft and having them remain stable in position despite the heating they would receive from the Sun.

A major advantage of the Mariner Jupiter/Saturn spacecraft over the Pioneer spacecraft was its sophisticated imaging system which would make available high resolution images of the planets and their satellites of the type obtained by earlier Mariners at Mars. The complex television-type imaging system, developed from those used on the Mariners to map the surface of Mars, was capable of revealing fine details on the surfaces of the satellites and intricate cloud structures on Jupiter and Saturn that were impossible to resolve with the relatively simple spin-scan imaging system of the Pioneers.

It is a tremendous technical challenge to obtain a high resolution picture from a spacecraft over half a billion miles from Earth. The field of view of the narrow-angle, high-magnification camera is one-third of a degree. It can be visualized by holding a soda straw to your eye and sighting through it. You see about one-third of a degree only. And the fields of view have to overlap slightly, but not more than 10 percent. Underlaps have to be avoided because these would cause parts of the surface being imaged to be missed. Too much overlap would result in requiring too many pictures when there is a time constraint as the spacecraft flies past a target such as a satellite. The satellite has its own motion, it is rotating on its axis, and it is revolving about the parent planet. The spacecraft is also moving, and all these things taking place simultaneously create a major problem in directing the cameras and arranging for the slightly overlapping pictures to be taken so as to cover all the visible surface of the satellite.

Remote sensing instruments, cameras, and infrared and ultraviolet sensors, had to be capable of being directed to look at satellites and planetary cloud features. Unlike Pioneer, but as on the Mariners to Mars, these instruments were mounted on a movable platform (figure 4-3) so that they could be pointed at targets irrespective of the orientation of the spacecraft.

Scan platforms had been used on Mariner. The problem with Voyager was in respect to slew rates (how fast the platform is moved) and stepping intervals (the amount of each movement) because many different targets had to be looked at during the relatively

Figure 4-3: The Voyager spacecraft carried a movable scan platform on which optical instruments were mounted. This permitted these instruments to be directed at targets on planets and their satellites as the spacecraft hurtled past them. Narrow- and wide-angle television cameras and a photopolarimeter are on the left. An infrared spectrometer-radiometer, with its large mirror, is at the bottom, and an ultraviolet spectrometer is on the right. Black colored milar blankets insulate the equipment. (JPL)

brief period of the flyby. Each time the platform is moved to direct its instruments to a specific target, this movement torques the spacecraft and results in acceleration which causes smearing of the image. To avoid this effect a time has to be allowed following each move of the scan platform for the spacecraft to settle down before a picture is taken. The whole system of step size, slew rate, and timing between the move and taking a picture, was a much more difficult problem than ever before.

Every picture required a separate set of commands to the scan platform. There was the challenge of not only making the platform so that it was controllable, but also defining the sequences and commanding all the instructions for these sequences to the scan platform.

Communications also had to be improved over those of the earlier Mariner spacecraft. A large 11.8-foot-diameter parabolic antenna was carried by the spacecraft. Also the frequency was increased to X-band in addition to the S-band communications used on the Mariners. These two changes permitted effective communications over the great distances to the outer planets in spite of the restrictions in the power of transmitters carried by the spacecraft.

The big antenna system for Mariner Jupiter/Saturn was new; the largest antenna flown in a planetary spacecraft. It was fabricated of honeycomb fiberglass, and was similar in design to the Viking Orbiter antenna. The first antenna stuck to the mold and destroyed it, with the consequence that another mold had to be made. This caused anxiety because of the long time it takes to make a mold of this kind. There were concerns that it could not be done in time for the launch. Also there were problems with the dual-frequency feed required for the antenna. The X-band feed was by direct illumination of the antenna, and the S-band feed was the secondary reflector. So the reflector had to be made transparent at one frequency and reflective at the other. The design approach was to embed small dipoles in a dielectric, but this resulted in a surface that was largely dielectric, and when the nature of an electrostatic discharge problem was understood following Pioneer's encounter with Jupiter, it was too late to do anything about the fundamental nature of the reflective surface. The consequence was that the antenna feed system and reflector became one of the biggest concerns for problems of electrostatic discharge. As a partial fix, some of the cables that went into that area for temperature measurements were eliminated.

The Mariner Jupiter/Saturn mission made use of the rare alignment of the planets in the late 1970s that had formed the basis for the Grand Tour (figure 4-4). The location of Jupiter relative to Saturn permitted a spacecraft to fly by both planets with only the launch energy needed for a mission to Jupiter alone. The year 1977 was the most desirable time for the launch. If the spacecraft were launched during that opportunity it could fly by Jupiter at a distance of 5 Jupiter radii, sufficiently far from the planet to

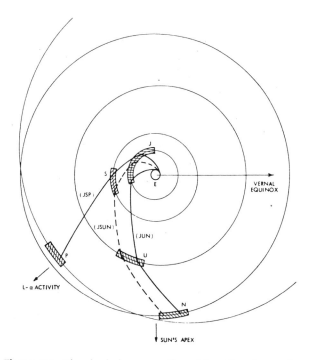

Figure 4-4: The shaded areas on the orbits of the planets show their positions in the period when spacecraft, launched between 1976 and 1980, could reach them using Jupiter and the other planets as gravitational slingshots.

(After P.A. Penzo, JPL)

protect it from radiation damage, and yet still close enough to obtain a suitable slingshot effect from Jupiter for the spacecraft to continue to Saturn in a reasonable time, and also close enough to Jupiter to obtain good images of the planet and its major satellites.

Although officially played down at the time, the line-up of the planets would also permit one of the spacecraft—the second to arrive at Saturn—to travel from Saturn to Uranus and Neptune. If all went well with the spacecraft it would reach Uranus in January 1986 and Neptune in 1989.

In May 1975, NASA issued an announcement of opportunity for scientists to participate in a Mariner Jupiter/Uranus mission being planned for a 1979 launch. But this project was later a victim of budget cutting and never materialized. Renewed interest then burgeoned about the potential of Mariner Jupiter/Saturn to reach Uranus.

Mission and system design was completed by the end of fiscal 1975 and all subsystem fabrication and tests had been completed by the beginning of fiscal 1977. Assembly of a proof test model started in July 1976 and the two flight spacecraft were to be completed by the end of fiscal 1977 when the launches would take place. At the time of launch the program cost just over $200 million, much less than the almost $1 billion required for the earlier Grand Tour missions. And the exciting thing was that those planning the mission knew that they could, if all went well, direct one of the spacecraft to visit all the outer planets except Pluto.

In April 1966, John R. Casani, who was then managing JPL's Guidance and Control Division after his outstanding management achievements on the Mariner Mars, and Mariner Venus/Mercury projects, was appointed manager of the Mariner Jupiter/Saturn project. He replaced H.M. Schurmeier, who had managed the program since its inception in 1972, as Schurmeier was to head the Civil Systems Division of the Laboratory.

Casani managed the project through building, testing, and assembly of the spacecraft and continued through the first and second launches. Then Project Galileo, the Jupiter Orbiter/Probe, was started and Mariner Jupiter/Saturn and Galileo were combined within a single outer planets office at the Jet Propulsion Laboratory. Casani managed this office, Ray Heacock was deputy for Mariner Jupiter/Saturn, and Al Wolfe was deputy for Galileo. That organization lasted for a couple of months only because too much was involved on the two programs to have an effective management structure of that type. So Mariner Jupiter/Saturn was separated again, with Heacock as manager for a short while until Robert Parks was appointed manager to carry the project through the first encounter.

Casani recounts some of the challenges while he was project manager. In addition to the usual problems and challenges of getting hardware through the final stages of design, there were several new and different aspects to Mariner Jupiter/Saturn: "This was the first time that we had attempted a computer-processor-based attitude-control system; previously these systems had been all hard-wired. Software development to support this approach was quite a challenge for us. We had some difficult times with the radio transmitter equipment . . . with traveling wave tube amplifiers. Each spacecraft carries four of them; two S-band and two X-band. The development problems with the tubes and the amplifiers was a major challenge for the project during the two years before launch. We were pushing the state of the art in two directions. We were trying to increase efficiency, in terms of direct-current input to radio-frequency output, and at the same time we were trying to improve the packaging density; in other words, trying for more amplifier in a smaller package for less power."

The most far-reaching challenge was one that permeated almost every design. It was the radiation problem at Jupiter. The first realistic assessment of the Jupiter environment was obtained from an understanding of the Pioneer data. It then became quite apparent that the Mariner Jupiter/Saturn spacecraft would have to be hardened (protected from radiation damage). The problem with radiation was particularly challenging because it had to be tackled after the design concepts for Mariner Jupiter/Saturn had been established.

"The designs were already fairly advanced as unhardened," explained Casani. "We had to go through them as well as we could and harden them. We had to substitute sensitive parts with others that were less sensitive to radiation. We went through the circuit designs and changed the designs to make them less sensitive to radiation effects, and we introduced shielding. This, however, made the spacecraft heavier. And even after taking all these steps there were still areas where the system was still too sensitive, so we went in and introduced spot shielding techniques where we selectively placed titanium or tungsten shields around individual components or pieces. The radiation concern also created problems of mission design in respect to selecting the trajectory for the flyby of Jupiter."

If the first spacecraft was sent past Jupiter within the orbit of Io to try to investigate the Io flux tube, there was a big risk of its being damaged by radiation. But the science return promised to be good. The second spacecraft would be directed to fly by Jupiter at 9.5 to 10 radii of the planet—between the orbits of Ganymede and Europa—where the radiation environment was a lot less hazardous. In that way all the project's eggs were not placed in the Jupiter science return basket, nor was everything ultraconservative.

The Canopus tracker—used to define the orientation of the spacecraft—was a problem in that it used operational amplifiers that were sensitive to radiation. Circuits had to be redesigned, but there was still the question of the Canopus trackers being basically photon multipliers. A radiation environment causes interference by producing secondary electrons in the tubes.

There are two concerns about radiation; one is the total dosage which causes permanent or semipermanent damage with only very slow recovery. The other is the interference effect, which is related to the flux of radiation and affects the gathering of data. All the hardware had to be requalified for the mission.

The computers carried by the spacecraft were also difficult to protect from radiation. Some components that could not be hardened were protected by being placed in steel containers or mounted on thicker structures that would absorb the radiation.

Earlier spacecraft had complete flight plans stored in their memories. This could not be done with Mariner Jupiter/Saturn; the mission was complex and several different planets were to be encountered. Also, redundant receivers were required, and they needed computers to allow for sophisticated science activities and enough autonomy to abort to a safe mode in case of failure. But they still had to be able to communicate with Earth, sending engineering telemetry only.

There were three computing subsystems on each spacecraft; one for telemetry, another the central sequencer, and the third for attitude control. Each computer had a backup, so the spacecraft carried six computers in all. A big problem was interfacing the computers so that they could talk to each other and check on each other. This required the development of complex software. A justification for choosing software control of telemetry and attitude (different from earlier Mariners) was that the system could be adapted to other missions; for example, to the Mariner Jupiter/Uranus mission, which was then being planned. And when that fell through, the software permitted the Jupiter/Saturn mission to be expanded into a Jupiter/Saturn/Uranus/Neptune mission.

The number of computers and their software systems made test and integration of the spacecraft a major challenge. Not only did the spacecraft have a computer-based attitude control system, which was new, but also the flight data system was computer-based, as was the command sequencer. Each was also redundant. These systems talked to each other and tested each other for faults. There were 32,648 words of memory and six processors and about 60,000 individual electronic parts—transistors, resistors, integrated circuits, and the like. Viking Orbiter, by contrast, had only half this number.

Another problem which was not quite as serious as the radiation problem, because it was not so pervasive, was that of electrostatic discharge, resulting from electrostatic charging of the spacecraft by the particles of the Jovian magnetosphere. This problem surfaced very late in the program, in fact after the spacecraft was shipped to the Kennedy Space Center. As the spacecraft travels through the radiation belts it is flying through the plasma from which it collects electrons. Ideally, if the spacecraft is completely unipotential, as a metallic surface is, there would be no differential. The spacecraft might accumulate a charge relative to the local plasma but that by itself would not be too bad. However, if there are metallic surfaces next to a dielectric surface, or two metallic surfaces insulated from one another, differential charging occurs and this can reach several thousand volts. With potential differences of that magnitude there are opportunities for electrical discharge that can damage electronics or invalidate data. The solution was to install bonding (electrical connections between conductive surfaces), change surfaces that were dielectrics, and provide resistive paths for charges to dissipate between different surfaces in a reasonable time. All the shields had to be bonded, and cables that ran outside the spacecraft and alongside insulating structures, such as the main antenna boom that was made of glass filament, were of particular concern. Cables and surfaces had to be treated in a way that differed from original planning, to prevent charges building up and arcing into the cables. Tests were devised in which the spacecraft was charged to a high voltage, and unexpected sensitivities were discovered even after everything thought necessary had been done. The spacecraft required further treatment.

Temperature control on the spacecraft was another major challenge. In some respects it was easier than for spacecraft going in toward the Sun to Mercury, but there were problems. The antenna served as a parasol so there was no concern about heat input from the Sun. The problem was in heat generated internally within the spacecraft. But in the outer

Solar System where there is virtually no solar heating, heat must not be lost through uncontrollable leaks into space. Insulating blankets on the Mariner Jupiter/Saturn had to be more carefully engineered than anything done previously for spacecraft. On the Viking Orbiter spacecraft there were 50 or 60 thermal blankets. Mariner Jupiter/Saturn required nearly 250 blankets. In addition there were louvers to reject excess heat in a controllable way. And there were electrical heaters to replace the heat from equipment when turned off. There were also many radioisotope heaters—one-watt pellets of radioisotopes encapsulated in material and attached to areas not very well coupled thermally to the spacecraft, for example, the magnetometers on their extended booms.

Charles E. Kohlhase was mission analysis and engineering manager for the mission. He recounts how the Mariner Jupiter/Saturn trajectories to the outer planets (figure 4-5) were selected by an involved process of tradeoffs among many variables.

The capabilities of the Titan/Centaur launch vehicle provided sufficient energy to allow a 400-day period for the arrival at Saturn, where that planet and its huge satellite, Titan, were prime targets for the Mariner Jupiter/Saturn mission. There was a 30-day launch window and 400 possible arrival dates; therefore a total of 12,000 Earth-to-Saturn trajectories via Jupiter were available. The requirement of an encounter with Jupiter's fascinating satellite, Io, eliminated some of these trajectories. The requirement of a close approach to Titan also limited the choice of Saturn arrival dates. There were science debates about the pros and cons of approaching Titan before or after the encounter with Saturn. The geometry of the ring occultation also controlled when the encounter with Titan could take place. The window for this was only plus or minus 30 minutes every 15 days (the orbital period of Titan). So 15-day, 30-minute slots defined the encounter further. Then cancellation of the Mariner Jupiter/Uranus mission made it necessary to try to find out if Mariner Jupiter/Saturn could go to Uranus. This again narrowed the choice of trajectories.

Arranging for passage outside the rings of Saturn, and trying to approach as many satellites of Saturn

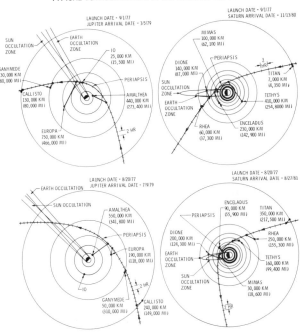

TYPICAL VOYAGER FLYBYS OF JUPITER AND SATURN

Figure 4-5: To decide on trajectories for the two Voyagers, mission planners had to eliminate thousands of possibilities. This was done by selecting trajectories that during encounters with Jupiter and Saturn would provide close encounters with specific satellites as well as chosen orientations of the paths past the planets. The top drawings show encounter trajectories with Jupiter and Saturn for a launch date of September 1, 1977, the bottom two are for a launch date of August 20, 1977. (JPL)

as possible and to arrange to see both hemispheres of each Galilean satellite, narrowed the choice still more. Finally, the 12,000 possible trajectories were reduced to a mere 98, for which it became practical to compute launch guidance sets and have them available for the spacecraft.

In March 1977, only a few months before the first launch, the name of the project was changed from Mariner Jupiter/Saturn to Voyager. The mission to the giant outer planets consisted of two spacecraft launched in August and September 1977 to reach Jupiter in 1979 and Saturn in 1980 and 1981 (figure 4-6). The project is managed by the Jet Propulsion Laboratory, which was responsible for building the two spacecraft and for conducting tracking, com-

VOYAGER FLIGHT PATHS

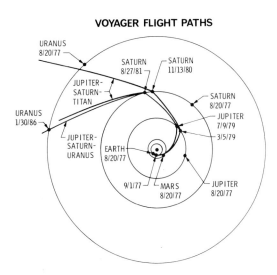

Figure 4-6: Voyager flight paths, showing how the first spacecraft to be launched in August 1977 would reach Jupiter and Saturn after the second spacecraft (launched in September) and would encounter Uranus in its orbit in January 1986.

(JPL)

munications, and mission operations. NASA's Lewis Research Center, Cleveland, Ohio, had responsibility for the two Titan/Centaur launch vehicles (figure 4-7).

Voyager became the most far-reaching space mission to be flown by NASA in that one of the spacecraft has the potential of visiting four different planets, the final visit to Neptune in 1989 being at a distance of 2.8 billion miles—30 times Earth's distance—from the Sun.

The Voyager Spacecraft

Each Voyager spacecraft (figure 4-8), like the Mariners, was built around a group of structural bays which house the electronics. Supported by a tubular trusswork on top of the bays is the big 12-foot-diameter, high-gain antenna. This antenna is pointed toward Earth by use of Sun sensors on the spacecraft. At the great distances of the outer Solar System, the Earth always appears close to the Sun as viewed from the spacecraft. So the antenna can be pointed toward Earth by offsetting it a few degrees only from the Sun.

In the center of the bays is a large spherical tank of hydrazine. Unlike earlier Mariners, the new spacecraft does not carry a main rocket engine for trajectory corrections: instead it uses, like Pioneer, small hot gas thrusters—16 in all for Voyager. These supply thrust for attitude control and for maneuvering. The spacecraft can be oriented by reference to the Sun and to the star Canopus, or by its own internal system of gyroscopes in an inertial reference unit.

Figure 4-7: Standing 15 stories high and weighing almost 700 tons, the Titan/Centaur launch vehicle was used to send the Voyager spacecraft into the outer Solar System. It consisted of a liquid propellant rocket core and two solid propellant motors main stage, and the high-energy Centaur upper stage. The spacecraft was enclosed within the protective nose fairing, the white conical structure on top of the launch vehicle.   (JPL)

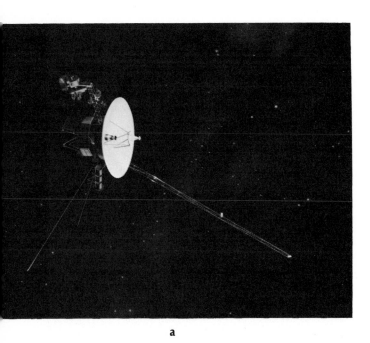

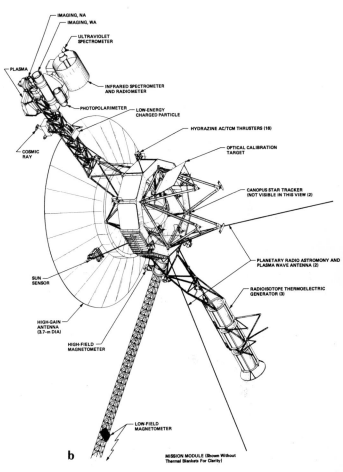

Figure 4-8: The Voyager spacecraft.

(a) Artist's concept showing Voyager in its interplanetary cruise configuration. (JPL)

(b) Drawing to identify the main subsystems of the spacecraft. (JPL)

For some of the trajectory corrections the small thrusters had to operate for as long as one hour.

Because the energy required to achieve a Jupiter trajectory with the large payload of Voyager was greater than could be supplied by the Titan/Centaur launch vehicle, each Voyager needed an additional propulsive stage. It consisted of a solid propellant rocket motor in a propulsive module which was detached from the spacecraft after use. The rocket motor was ignited 15 seconds after Voyager separated from the Centaur. It developed a thrust of 16,000 lbs for about 43 seconds to add about 4,475 mph to the speed of the spacecraft. About 10 minutes later the whole propulsion unit was jettisoned.

Within the electronics bays several of the subsystems—such as command, control, and sequencing computer subsystems and the flight data subsystem—are duplicated to provide backup in case of a failure of these units which are so essential to the continuance of the mission.

The whole spacecraft was designed to be as magnetically clean as possible so that scientific measurements would be affected minimally by the magnetic field of the spacecraft itself.

The power system is an array of three radioisotope thermoelectric generators shown on their boom in figure 4-9. These RTGs, mounted in tandem on this

Figure 4-9: The size of the Voyager spacecraft is apparent from this photograph, which shows the three radioisotope thermoelectric generators on the left. The model is pointing to the message to extraterrestrials discussed in the Epilog.   (JPL)

memories. Half of each memory stores fixed computer instructions needed throughout the mission. The other half stores commands from Earth. For some commands which could cause serious mission problems if given incorrectly—for example, commands for a trajectory correction maneuever—the identical command has to be given by both the processors of the system before it is implemented. The second computer-based subsystem is for attitude control and propulsion. It consists of hybrid programmable attitude control electronics, redundant Sun sensors, redundant Canopus star trackers, three two-axis gyros, and scan actuators to position the scan platform. It, too, has two redundant memories, part of which is programmable and the other part fixed.

The third computing subsystem was the programmable flight data subsystem which controls the science instruments and formats all science and engineering data for transmission to Earth. The spacecraft can transmit to Earth at two frequencies, S-band (approximately 2295 MHz) and X-band (approximately 8418 MHz). Signals from Earth to the spacecraft are at S-band only. During cruise the lower frequency S-band—as used on the earlier Mariner spacecraft—is used to send data at 2,560 bits per second. This is adequate for cruise science and releases the big antennas of the Deep Space Network for other tasks. S-band can be received at this bit rate by the smaller antennas of the Deep Space Network. For encounters with the planets, where voluminous amounts of data must be transmitted quickly, the X-band is used. The S-band then acts as a dual frequency transmitter for those radio experiments where precision is considerably enhanced by simultaneous observation at two frequencies. The X-band transmitter operates at 12 or at 21.3 watts; the S-band at 9.4 or 28.3 watts, but they are not operated simultaneously at high power. Both are duplicated in case a transmitter should fail during the long mission.

Data transmitted from the spacecraft can be sent at either high or low rates. The low rate operates at S-band at 40 bits of information per second and is used exclusively for sending engineering data during pla-

boom that was deployed after the spacecraft separated from the launch vehicle, convert into electricity heat released by the decay of plutonium-238. For the mission to Jupiter and Saturn they produced a power output of 480 watts at the beginning of the mission. The output gradually decayed to about 385 watts by the time the spacecraft passed Saturn. A more long-lasting set of units was required for the mission to Uranus and Neptune.

Because the spacecraft have to operate for so long in space, charged capacitor energy storage banks are used instead of batteries to supply extra power when for short periods the spacecraft requires more power than the RTGs can supply. Excess power from the constant output RTGs is dumped overboard to space through a heater element outside the spacecraft.

The heart of the onboard control system of the spacecraft is the computer command subsystem mentioned earlier. It has two independant electronic

netary encounters when data requiring a high rate of transfer are transmitted at X-band. The high rate channel is used throughout the mission on either S- or X-band. It can accommodate engineering data at 40 bps (bits per second) or 1,200 bps, and real-time (transmitted as gathered) cruise science and engineering data at 2,560, 1,280, 640, 320, and 80 bps. If needed, slower rates of 40, 20, and 10 bps can be used at great distances beyond Saturn.

At Jupiter real-time general science and engineering data were transmitted at 7,700 bps and, for brief periods only, some science data were transmitted at 115,200 bps. Also, real-time data for general science, engineering, and television during encounter were transmitted at 115,200, 89,600, 67,200, 44,800, 29,966, and 19,200 bps, and recorded data at 67,200, 44,800, 29,666, 21,600, and 7,200 bps. All the above rates are available at X-band only.

Additionally, memory data stored in the three computing subsystems can be read out and played back at 40 or 1,200 bps at either S- or X-band.

This great variety of data transmission capability is required because of the enormous changes in the length of the communications link to Earth, which has to extend as far as Neptune for the second Voyager, and because of uncertainties about the amount of interference that might be encountered on the signals.

Science Instruments

Originally the mission was concerned primarily with the two giant planets Jupiter and Saturn and a comparison of their physical characteristics. However, increasing interest in the satellite systems and the possibility of there being unusual development of these planet-sized bodies led to an expansion of mission objectives to include the satellites. As already mentioned, the science experiments are quite extensive in their scope. Cameras image the planets and satellites to a detail never before possible. Other instruments investigate the unusual particles and

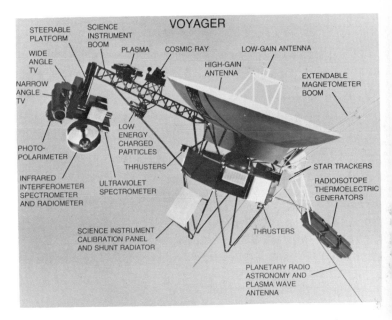

Figure 4-10: All the science instruments are identified on this view of Voyager.   (JPL)

fields environments of these big planets. Many of these experiments are identified in figure 4-10, and they can be grouped somewhat similarly to the Pioneer experiments.

Particles and Fields

A boom, extending 42.65 ft from the spacecraft, carries two magnetometers to measure low-intensity magnetic fields. They are spaced at different distances from the spacecraft so that the effects of the spacecraft can be eliminated from their measurements. Two magnetometers capable of measuring high-intensity fields are mounted on the spacecraft close to the base of the boom. The magnetometers are of triaxial fluxgate design. The low field magnetometers measured fields ranging from 0 to 50,000 gamma (to 0.5 gauss) to detect the fields in interplanetary space. For the more intense field of Jupiter, the high field magnetometers covered the range of 12 to 2,000,000 gamma (to 20 gauss).

On the science boom carrying the articulated scan platform, but not on the scan platform itself, several other instruments are mounted; a plasma detector, cosmic ray detectors, and a low-energy charged particle detector.

Dual Faraday-cup plasma detectors, one pointed toward the Sun and the other pointed 90 degrees from the Sun, record particles with energies between 10 eV and 6 keV, and 4 eV and 6 keV respectively. The instrument was used to detect hot subsonic plasma in the planet's magnetosheath and cold supersonic plasma in the solar wind.

Information on the types of energetic particles and their direction of flow is provided by a set of detectors covering the energy range between 10 keV and 30 MeV. A low-energy charged particle instrument is used primarily to detect particles in the magnetosphere and from the Sun. High-energy particles in interplanetary space are detected by a system of solid-state cosmic ray detectors for a range from 0.15 to 500 MeV. New detectors developed for Voyager were capable of obtaining more information than was possible with the instruments carried by the Pioneers.

Measurement of high-speed (relativistic) electrons with the cosmic ray system, coupled with plasma, low-energy particles, and magnetic field measurements, were intended to help improve our understanding of the basic plasma physics that permits electrons to be accelerated to such high velocities, as discovered by the Pioneers. Composition of the radiation belts of Jupiter (and later of Saturn) provided detailed information about their structure and mechanics. The magnetic moment of Jupiter was also determined, from which the internal structure of the planet was inferred.

The Jovian satellites (and the Saturnian satellites and ring system) offer obstacles to plasma fields co-rotating with the planet, analogous to the inner planets offering obstacles to the solar wind. The interactions of satellites with the planetary plasmas discovered by the Pioneers was investigated. Close approaches to Io, Ganymede, and Europa, permitted the interactions to be determined from radiation fluxes in their wakes. These science measurements were intended to throw light on how quickly the satellites sweep up particles trapped in the Jovian magnetosphere, and how soon the wake is refilled by inward diffusion of particles toward Jupiter.

Another experiment was to find out how much the spacecraft itself sweeps particles and becomes charged as it passes through a planetary magnetosphere. Some Earth orbiting artificial satellites accumulated large potentials, and observations suggested that the Pioneers became charged to a high potential as they flew by Jupiter, leading to modifications of the Voyager spacecraft (as described earlier) to protect them from high voltage discharges.

Optical Instruments

The other major group of experiments was concerned with the instruments carried on the scan platform. The atmosphere of Jupiter and the surfaces of the Galilean satellites were probed by these instruments which, together with radio science, considerably increased our understanding of the outer planets. Planetary atmospheres are important because they tell the story of the way in which the planets might have evolved since their formation, and also provide clues to the formation itself. Indeed, any theory about the formation and evolution of the Solar System must account for today's differences in the atmospheres of the planets. Detailed knowledge about the present states and compositions of planetary atmospheres is vital to an understanding of how the Solar System originated and subsequently evolved to the present.

Scientists needed to know about the temperature, pressure, density, and gaseous and particulate composition of Jupiter's atmosphere. It has been found that internal heat plays a major role in the dynamics of this atmosphere. The same is true of Saturn and may also be true for the more distant planets. Voyager experiments were directed toward clarifying the types of activity in the atmospheres of the outer planets. In addition, the mission searched for and looked at atmospheres of satellites which might provide important clues about the evolution of those bodies and the planetary systems of which they form a part.

The imaging system consists of advanced television cameras mounted on a remote sensing platform at the end of a short science boom. This platform also carries an ultraviolet spectrometer, an infrared spectrometer and radiometer, and a photopolarimeter. A narrow-angle (telephoto) camera of 59-inch focal length uses essentially the same type of optical system as that of Mariner 10 for the mission to Mercury, but its electronics were improved. Each image contains 5 million bits of information, and resolves details to 4 seconds of arc. A completely new wide-angle camera was designed for the Voyager mission. It has a 7.87-inch focal length telescope. Both camera systems are equipped with eight filters, some of which are used to produce images that are integrated into color pictures of the planets and their satellites. Better color renditions were achieved than with Pioneer because of the ability to obtain images in three colors rather than two.

The infrared interferometer spectrometer used a 20.1-inch-diameter mirror telescope to measure temperatures at various depths in Jupiter's atmosphere and its composition.

The ultraviolet spectrometer operates over the range from 500 to 1,700 angstroms beyond the visible spectrum. It measured the abundances of ions, atoms, and small molecules of gases in the atmosphere of Jupiter. It was particularly useful in searching for the presence of hydrogen and helium, and their relative proportions.

The photopolarimeter is a variable-aperture telescope system with filters and polarization analyzers. It provided information about aerosols in the planet's atmosphere and the surface characteristic of satellites.

Observations for several weeks while approaching each of the outer planets provided records of atmospheric motions from which time-lapse motion pictures were constructed of the swiftly rotating atmospheres. This capability, while producing revealing pictures of the cloud motions in the Jovian atmosphere is especially important for Saturn, Uranus, and Neptune, because even coarse details of cloud patterns on these very distant planets are almost impossible to obtain from Earth-based observations.

Far encounter pictures of satellites using the high resolution camera system revealed details on their surfaces only 3 to 9 miles across. All four of the Galilean satellites were targeted for close approaches by the spacecraft to show details as small as 1,600 feet across. The cameras sought geological details of the satellites' surfaces—craters, plains, scarps, mountains, and polar caps. The lower resolution pictures revealed global distribution of provinces and showed why there are variations of color and albedo on the satellites.

Radio Experiments

Two long antennas project from the spacecraft to detect radio waves from the planets and plasma waves in the interplanetary medium. These 32.8-foot-long whip antennas of the planetary radio astronomy experiment were used with a stepped frequency radio receiver (scanner) to detect planetary radio emissions or bursts of radio waves over the frequency range 20.4 to 1,345 kHz and 1.23 to 40.55 MHz. The experiment also looks for radio emissions from the Sun and the stars. The plasma wave experiment uses the same antennas with a step frequency detector and a waveform analyzer to measure plasma waves in the interplanetary medium. It produced thermal plasma density profiles at Jupiter and revealed interactions of satellites with the planetary magnetosphere.

As with the Pioneers, radio science investigations used the spacecraft's radio communications to observe the effects on the signals when the spacecraft passed behind Jupiter. Voyager had the great advantage of being able to compare two radio frequencies

(X- and S-band). These observations provided information about the dimensions of Jupiter and its atmosphere. Additionally the signals were used to determine with greater precision the gravitational fields and masses of the planetary bodies and their positions in space, thereby improving their ephemerides. These radio experiments also benefitted from being made at a precise frequency generated by a new ultrastable oscillator carried by the Voyager spacecraft.

Some of the difficulties in developing the science experiments for Voyager were recalled by project scientist Ed Stone: "The most difficult science experiment to develop did not in fact get developed. It was the modified IRIS [infrared interferometer spectrometer] experiment. Around 1975 the Space Science Board's Committee on Planetary and Lunar Exploration had recommended a Uranus mission as the number one priority, and NASA solicited proposals for a Mariner Jupiter/Uranus Mission, but it got lost at Headquarters in 1975 as they were trying to work out the budgets. At that point the result was to take a look at whether Voyager could go on to Uranus, and if it could, what instruments should it have. One of the things that was clear was that there would be a great advantage to having an infrared interferometer with an extended capability to shorter and longer wavelengths. R. Hanel had proposed such an instrument for Mariner Jupiter/Uranus, and that instrument was selected, . . . as a replacement instrument for the IRIS on Voyager. But it did not get finished in time. He did not start working on it until the beginning of 1976, and we launched in 1977. That was the most difficult instrument to make, and it was just not possible to do so in the time."

Of the other instruments, Stone said that the planetary radio astronomy experiment had not been flown before. The instrument had a totally new design. IRIS had been flown on Earth-orbiting satellites and on a Mariner. The ultraviolet spectrometer was a direct descendant of what L. Broadfoot had flown on Mariner 10. The photopolarimeter did not have a direct heritage with anything that had been done before, but it was a rather simple instrument.

The plasma instrument had an interesting development. "It was a Faraday cup but a much larger one than had ever been flown before. There were some technical problems in making it so big, but they were not fundamental problems. JPL [Jet Propulsion Laboratory] built the imaging system, then an imaging team was selected to use the system. JPL also built the radio system and there was a radio science team to use that system. For all nine other investigations the principal investigator had the responsibility of building the hardware and delivering it."

Launch of the First Voyager

The first spacecraft was being readied for launch in early August 1977 (figure 4-11), and launch was

Figure 4-11: The Voyager spacecraft is here being readied for its tremendous Voyage to Jupiter, Saturn, Uranus, and possibly to Neptune.   (JPL)

scheduled for the 20th, then the attitude and articulation control subsystem and the flight data subsystem failed under test. Because time was so short, a decision was made to exchange the two flight spacecraft. This also required that the RTGs be switched because those in the first spacecraft to be launched needed higher power to be capable of maintaining sufficient output until the spacecraft reached Uranus. The failed equipment was returned to the Jet Propulsion Laboratory in Pasadena.

At the Laboratory, mission operations continued to generate sequences and perform test and training exercises.

But there was another hold-up on August 9 after the spacecraft was placed within the protective nose shroud of the launch vehicle. During a test routine, engineers discovered that the low-energy charged particle instrument was not grounded. Consequently the spacecraft had to be removed from the shroud, and another grounded charged particle instrument installed in it. Back within the shroud again, the spacecraft successfully completed all tests, and was then mated with the launch vehicle. Everything was ready for the launch, and strenuous efforts were underway to make the other spacecraft ready for August 18 so that it could support the first launch if there was an aborted launching. But an intermittent condition was detected in the hardware of the backup spacecraft's telemetry system. So a telemetry system from the proof-test spacecraft had to be taken out and installed in the flight spacecraft. By this time the flight data system computer had been repaired at the Jet Propulsion Laboratory and flown to the launch site. It too was installed ready to backup the launch.

A full operational readiness test was completed successfully on August 15, and a practice countdown was completed. All the computers onboard the substitute spacecraft were now operating normally.

On August 18, the countdown went smoothly until five minutes before launch, when the status of a valve in the launch vehicle was in doubt. It appeared to be stuck and had to be checked. When this problem had been cleared, the countdown continued. Then, five minutes into the launch window, at precisely 10:29:45 A.M. EDT, Voyager 2 aboard a Titan IIIE/Centaur launch vehicle lifted majestically from launch complex 41 at Cape Canaveral, Florida on its way to Jupiter, Saturn, Uranus, and possibly Neptune, and then on to the stars (figure 4-12).

After the launch a batch of small problems surfaced within the spacecraft. There was a suspected gyro failure, incomplete data transmission, and uncertainty about the deployment of the science platform boom.

Software routines within the spacecraft detect faults in redundant subsystems, and if a fault is found the

Figure 4-12: At 10:29 A.M. EDT on August 20, 1977, Voyager 2, the first spacecraft to be launched, left Cape Canaveral on its voyage to the outer Solar System.   (JPL)

software then decides what has to be done about it in a programmed sequence of steps. These steps are designed to isolate what is wrong and to switch that part of the subsystem out and replace it with a backup part. This is a time-consuming logical process.

During the launch the gyros were saturated by a roll turn of the launch vehicle. This would, of course, not happen during typical flight through space. The saturated gyros were interpreted by the fault detection software as a failure. The program began to go though the sequence of switching gyroscopes in the spacecraft and even switched processors when the switched gyros still showed the same problem. The program was, in fact, doing exactly what it had been instructed to do; check a subsystem, switch to a redundant system, if the redundant system does not correct the problem, then check and switch processors. But the system became so tied up in following this routine during the launch that there was no processing time available for anything else, and expected telemetered data were not received on the ground. So when the spacecraft separated from the launch vehicle there was no processing time free and the attitude of the spacecraft was incorrect.

Said project manager Casani, "We could not tell on the ground what was going on. We thought we had a failure within the spacecraft. It was not until after the spacecraft separated from the third stage that we found there was not a real problem. These effects were unanticipated and it took us quite a while to figure out what had happened."

"You never really understand a spacecraft until you've flown it," comments Bud Schurmeier.

The boom problem was somewhat different. The science boom carrying the important scan platform with its instruments so vital to the mission was supposed to be released about 53 minutes into the flight, but telemetry data gave no indication that it had extended and locked into place. Engineers suspected a faulty microswitch that should have closed when the boom locked in its deployed position. Twelve hours after launch, the plasma science instrument

was turned on. This instrument is mounted on the scan platform and its data indicated that the science boom had, in fact, extended. But was it latched in place? In subsequent days a number of tests were made. A plan was developed to jettison the infrared interferometer spectrometer cover simultaneously with the spacecraft being rotated so as to torque the boom in the direction needed to latch it. However this maneuver had to be aborted when trouble developed in the attitude control system. Star fields were imaged to check if the scan platform was moving as it should. These indicated that the hinge on the boom was very close to being in a locked position, but not quite home. Finally engineers decided that the locking sensor was at fault and the boom was actually in place and operational.

Then the spacecraft was placed into a quiet mode so that everyone could concentrate on the launch of the second spacecraft at the beginning of September.

The Second Launch

To make sure that the science boom would lock correctly on the second spacecraft, the launch was delayed while engineers rapidly designed a better mechanism by installing five coil springs on the boom. These were intended to force the boom into a safe locked position. August 31 the spacecraft was mated with the launch vehicle. A practice countdown started two days later. The launch countdown went smoothly without any holds. On September 5, at 8:56:01 A.M. EST, Voyager 1 followed Voyager 2 into space (figure 4-13) on a journey to Jupiter and Saturn, and then on to the distant stars. The second spacecraft was named Voyager 1 because it would arrive at Jupiter first. The first spacecraft to be launched was named Voyager 2. Although launched 16 days after the first spacecraft, the second spacecraft followed a trajectory that would cause it to arrive at Jupiter four months ahead of the first spacecraft.

During the launch, the Titan stage II burned for less time than planned. The Centaur had to make up for

Figure 4-13: Just over two weeks later, at 8:56 A.M. EDT on September 5, Voyager 1 followed Voyager 2. The second spacecraft would reach Jupiter four months ahead of Voyager 2.   (JPL)

this by having a longer first burn, which in turn used 1,200 pounds more propellant than had been planned. Mission planner Kohlhase explained what happened: "The Titan lifted off fine. But then there was a cliff-hanger. Centaur had 1,200 pounds extra propellant to establish a parking orbit. However, a fuel mixture ratio valve on the Titan failed to give the Centaur enough energy. The big question then was would the Centaur itself burn short. It was close. Fortunately, the launch date was not at an energy barrier because there was a propellant reserve. Also the Centaur was lighter, so its acceleration was greater."

"Afterwards it was found that there were only 4 seconds of propellant left when the Centaur engines cut off. If it had happened on the first launch Voyager might not have made it. And if the second launch had slipped 7 days beyond 5 September due to weather, for example, that spacecraft would not have made it."

Voyager 1's first trajectory correction maneuver was made in two parts on September 11 and 13. The change in velocity obtained was, however, 20 percent less than expected for each part of the maneuver.

At a distance of 7.25 million miles from Earth, Voyager 1's camera recorded a unique picture of the Earth and Moon (figure 4-14) on September 18, 1977. The Moon was above and beyond Earth as viewed by Voyager. The illuminated crescent of Earth showed eastern Asia, the western Pacific Ocean, and part of the Arctic. In fact, the spacecraft was 7.25 million miles directly above Mount Everest, Earth's highest mountain, when the picture was taken. Because the Earth is many times brighter than the Moon, the Moon's image had to be computer enhanced by a factor of three so that both Earth and Moon could be shown on the same picture.

On September 23, Voyager 2 experienced a problem in the flight data system with a loss of some data of engineering measurements.

There was concern about both spacecraft by mid-October. Both had been consuming more propellant than anticipated. This resulted from impingement of the thruster jets on the struts, solar radiation pressure effects, reaction to the starting and stopping of the digital tape recorder, and motion of the scan platform. Charles Kohlhase again explained one of these problems: "The four thrusters that make the translational corrections, not the ones that rotate the attitude, are located on the back of the spacecraft and they all thrust in the same direction. There are some struts where it [the spacecraft] was attached to the launch vehicle. These struts are actually not just narrow metal struts but they are wrapped with thermal blankets and . . . if you could stand at one of the exhaust nozzles and look out you would not have a clear view of space. You would see a little bit of strut."

The designers had calculated that there would be an impingement of gas that would be contrary to the normal thrust direction. They had estimated the effect as 6 percent but it turned out to be much greater.

Figure 4-14: On September 18, 1977, Voyager 1 obtained this unique picture on the Earth/Moon system at a distance of 7,250,000 miles from Earth. The Moon is at the top of the picture and beyond the Earth as seen from Voyager. (JPL)

than made up for this effect. For Voyager 2 . . . we needed every bit of delta-vee to get it to Uranus, and we suddenly learned it was not going to deliver that amount."

After much analysis of the situation, mission planners discovered they could recover from the loss of delta-vee capability by changing a maneuver that would be needed after the spacecraft flew past Jupiter. The original mission plan was to make the necessary maneuver (to reach Saturn with the option of continuing to Uranus) at 70 days after the Jupiter encounter. But the longer the delay after leaving Jupiter and before the correction could be made, the more it would cost in propellant. So the solution was to pull back the maneuver closer to Jupiter. First the trajectories were calculated for a maneuver 14 days after Jupiter encounter. Then the maneuver was pulled back further to two days after encounter. Later still it was discovered that the cheapest and most reliable maneuver would be one made very shortly after the closest approach to Jupiter when the direction to Earth lay exactly along the trajectory of the spacecraft. An Earth-line burn, in which the direction of thrust pointed along the path of the spacecraft and directly away from Earth, did not require that the spacecraft be reoriented for the burn. It would slow Voyager 2 by the required amount and save propellant for the journey to Uranus. But this would still not save enough propellant to be sure of getting to Neptune after a Uranus encounter.

Kohlhase went on: "We were trying to slow the spacecraft to arrive [at Saturn] on the 27th of August [1981]. If we could have found an earlier arrival date that was still attractive then we would not have to slow up even that much. We knew that the two satellites [of Saturn] Voyager 1 would not come close to are Tethys and Enceladus. And we also knew of Voyager 2 that if we came back to about August 25 for the encounter, those moons would have orbited even numbers of revolutions and we could come close to Tethys and Enceladus, thereby complementing the Voyager 1 trajectory and also reducing the fuel further. So we changed the nominal arrival date from the prelaunch selection to the 25th of August.

"We had the two spacecraft under way and they could not supply the number of meters per second delta-vee [velocity change]. We asked: What about Voyager 1? No problem on it, because . . . Saturn is its last target and it is not going on to Uranus or Neptune. Also we launched 4 days late so we had extra pounds of propellant [hydrazine] which more

"But [although] that set us more fuel aside for going on to Uranus and Neptune it had one drawback. The original trajectory [of Voyager 2 to Saturn] was chosen to have a repeat close Titan encounter if Voyager 1 failed. The minute we changed the arrival date, Titan moved away from the good spot. So if we ever wanted to go to Titan again [with Voyager 2] instead of to Uranus, we would have to do a big maneuver to get back, sometime in January 1981. To go back to the close Titan encounter, the delta-vee would be 70 meters per sec. It would take most of the fuel we had."

Everything would rest upon the success of Voyager 1 obtaining a good encounter with Titan, the intriguing large satellite of Saturn with an unusual atmosphere that is denser than the atmosphere of Earth.

So if the required maneuvers were made at Jupiter, Voyager 2 would have enough propellant to reach Uranus. Nevertheless mission planners investigated several methods of conserving propellant supplies, including using the star Deneb instead of Canopus as a reference point, and reducing the sensitivity of the Sun sensors.

Voyager 2's photopolarimeter developed problems in its filter wheel, which stuck in one position. And in December 1977 Voyager 1's camera filter wheel also developed trouble. It was turned off, and when brought on again in February 1978 it operated normally. The problem was a bad memory location in the computer of the flight data system. This memory location was changed to a spare. And in March the stuck analyzer wheel of Voyager 2's photopolarimeter became unstuck, but the instrument itself behaved erratically, so it was switched off.

Also in February 1978 a routine cruise maneuver for Voyager 1 was interrupted when the computer of the command control system entered a failure protection mode. This resulted from the spacecraft's gyro reference system not working correctly. About this same time the plasma instrument developed a problem when its cluster of three detectors lost sensitivity. And the scan platform slowed to a standstill on February 23 and did not complete its commanded movement. Controllers issued commands for the scan platform to be moved in low gear, thereby applying the most torque. Commanded to back off from its stuck position, it moved on March 17, but not as far as expected. Then two more moves were commanded and both worked. Six days later it was moved again four times and project officials hoped that all would be well for the Jupiter encounter.

In early April the main radio receiver of Voyager 2 failed. The onboard computer automatically switched to the backup receiver. A faulty tracking loop in the backup receiver compounded the problem of communications. Attempts to obtain two-way lock on the backup receiver failed. The computer switched back to the primary receiver when no communication had been received from Earth for 12 hours. Then the main receiver seemed to be operating normally. But near disaster struck. An excessive current in the receiver blew the fuses and communication with Earth was broken again; the spacecraft was unable to receive any commands. The controllers now had to wait seven days for an automatic timer to switch in the second receiver again.

On April 13, shortly after 3:30 A.M. a command was sent from the Madrid Station of the Deep Space Network to Voyager 2. Everyone waited anxiously to see if the spacecraft would acknowledge its receipt. Nearly one hour later a signal came back from the spacecraft confirming that the command had been received. But although the receiver was operable it could not follow a radio signal whose frequency was changing. And signals were changing in frequency because of Doppler effects resulting from Earth's rotation. To overcome this problem, engineers calculated how much a signal would be changed in frequency by the Doppler effect, and then they adjusted the frequency of the transmitter to compensate for this so that the spacecraft would receive an unchanging frequency signal. First, however, the engineers had to find out the exact frequency to which the receiver of the spacecraft was most sensitive. They did this by transmitting a command sequence over a rising frequency range.

A command was also sent to stop the X-band transmitter from turning on. This was done to keep temperature within the spacecraft stable, because changing temperature could affect the receiver's frequency. Another command switched on the S-band transmitter at full power so that the spacecraft could be tracked by the smaller 85-foot diameter antennas of the Deep Space Network. During the communications emergency all the big 210-foot-diameter antennas had been dedicated to tracking the ailing spacecraft. This meant that other projects—Viking, Pioneer, and Helios—had to forego use of the big antennas for a period, until communication with Voyager was again assured.

In May the plasma instrument on Voyager 1 came back into operation during a series of tests to correct what was believed to be a problem in an amplifier. On the way spacecraft sometimes appear to have minds of their own. Schurmeier once commented: "We don't send them commands; we sent them requests."

Meanwhile, scientists and project management were concerned about the effects of the command receiver failure on Voyager 2, now that the primary receiver was dead. They developed a contingency plan in which a backup mission sequence was commanded to the spacecraft and stored in its memory. Should the remaining receiver fail, the spacecraft would at least have a science program that it would be able to follow during the Jovian and Saturnian encounters, and all would not be lost. The sequence included instructions for the 11 science instruments, including imaging at Saturn, but not at Jupiter. The commands also included instructions for executing a trajectory correction maneuver after passing Jupiter that would ensure the spacecraft would reach Saturn.

In July 1978 both spacecraft passed on the far side of the Sun away from the Earth and they were virtually shut down for this period.

One year after its launch, Voyager 1 was 449 million miles from Earth, on September 5, 1978. One year

Figure 4-15: Soon the Voyagers were approaching Jupiter and producing pictures of unsurpassed beauty as the high-resolution television cameras started to fill in all the intricate detail suggested by the earlier pictures from the Pioneers. The true majesty of the giant planet of the Solar System was about to be seen for the first time.   (JPL)

after Voyager 2's launch, that spacecraft was 431 million miles from Earth. By this time the problem with the scan platform of Voyager 1 had been resolved, and no difficulties were being encountered. It was deduced that the problem had been caused by bits of plastic debris blocking the gears and this debris had been crushed so that it no longer blocked the movement.

The two spacecraft cleared the asteroid belt safely; Voyager 1 on September 8, and Voyager 2 on October 21, 1978, and in January 1979, when Voyager 1 was 37 million miles from Jupiter, its observatory phase of operations began (figure 4-15). The Voyagers were close to their encounters with the giant.

5.

Close Looks at Jupiter

JANUARY 4, 1979: despite minor problems experienced at the beginning of the mission, Voyager 1 behaved admirably as it approached Jupiter. Sixty days and 38 million miles from Jupiter, all equipment was operating as it should at the beginning of the observatory phase of the mission with the spacecraft rushing toward the Jovian system. Voyager's pictures of Jupiter now began to rival the best taken from Earth. Already hurtling through space at 30,000 mph, the spacecraft was gradually picking up speed. But it would be another two months until Voyager would skim by the giant planet only 130,000 miles above the variegated cloud tops.

Every day, Voyager returned pictures of Jupiter showing ever-increasing detail. Within a few weeks details began to be revealed on the Galilean satellites too. They changed from points of light (figure 5-1) to interesting worlds. By January a picture of Jupiter was obtained that showed Io moving across the face

of the planet and revealed the dark poles of Io and a bright region near the equator. Pictures of the Great Red Spot soon confirmed its counterclockwise circulation and anticyclonic nature (figure 5-2). That it was a raised system of cloud some five miles above the surrounding clouds of ammonia crystals seemed indisputable.

On February 10 Voyager entered the Jovian system when it crossed the orbit of the outermost known satellite, Sinope, which is believed to be a captured asteroid. Sufficient pictures had been received of Jupiter to make a time lapse motion picture of the planet. To the excitement of meteorologists, this display of planetary dynamics revealed for the first time the complexity of the cloud motions and the jet streams swirling around the giant of the Solar System. The meteorologists saw motions never before revealed in the Jovian belts and zones. They observed the curious phenomenon of dark orange colored hot

Figure 5-1: Voyager 1's camera captured two of Jupiter's moons, Ganymede (right) and Europa (top right) on the morning of January 19, 1979, from a distance of 29 million miles. Details are beginning to show on the surfaces of these satellites. The brightness of Europa compared with Ganymede is clear. But this picture shows the dark hemisphere of Ganymede, its other hemisphere is brighter. Although Europa is believed to be a rocky satellite its surface appears to be covered with ice. Europa is about the size of Earth's Moon, Ganymede is slightly larger than the planet Mercury.   (JPL)

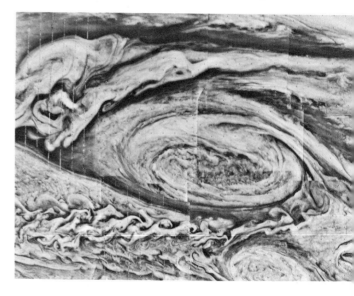

Figure 5-2: This mosaic of the Great Red Spot is made up of 12 pictures from Voyager 1 taken March 4 at a distance of 1,800,000 miles. Although the periphery of the Red Spot shows cloud streaming, which is characteristic of its counter-clockwise circulation, the cloud patterns near the center suggest substantially reduced vorticity. The smallest clouds visible in this picture are about 20 miles across.   (JPL)

spots moving through the upper layers of the clouds. These spots sometimes overtake one another and merge into one, and mysteriously break up again days later.

Striking changes were also revealed in the Great Red Spot (figure 5-3). Since the flybys of the Pioneers, the color of the spot had changed from orange to a dark brown. And now there were swirling atmospheric currents around the spot that were not present in 1973 and 1974. But the nature of the heat source required to drive the atmospheric mass of the Red Spot so high above its surrounding was still much in doubt. The spot has been located at one latitude, 20 degrees south, since its discovery three centuries ago. But it migrates in longitude at varying speeds. Observations from Earth also suggested a 90-day periodicity or oscillation in its east/west movements. Astronomers suggest that oppositely flowing belts

north and south of the spot may keep it rolling, but this does not explain why it has lasted so long.

A final trajectory correction maneuver was made on February 20. The hydrazine thrusters fired for 2.25 minutes to tune the path of the spacecraft so that it would fly past the giant planet precisely as required for all the science experiments.

At 3.8 million miles from Jupiter Voyager 1 passed through the bowshock of the planet on February 28, only to be overtaken a few hours later, as was Pioneer, by the collapsing shock as the solar wind pushed it in toward Jupiter. A short while passed before Voyager again penetrated the shock. And on March 1 at 5:00 A.M. another crossing took place, making five in all. But it was not until noon on March 1 that Voyager crossed the magnetopause to enter the magnetosphere of Jupiter. At a press conference about this time, project scientist Edward C. Stone

a

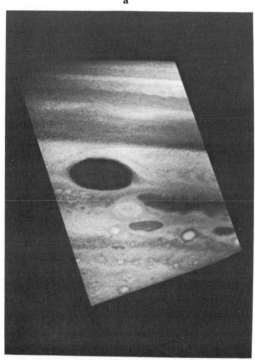

PIONEER 11   UNIV ARIZ
RANGE   640000 KM   PHASE 70   LIME   80
MID TIME OF DATA RECEIPT   9 DEC 00 30 UT
CB   BLUE   DATE 12 4 74

b

Figure 5-3: Changes in the Red Spot and its environs are apparent when these two pictures are compared.

(a) Taken by Pioneer 11 on December 4, 1974, at a distance of 338,000 miles. Note the dark and light ovals below and to the right of the spot. The bright ovals first formed about 40 years ago, and they move relative to the spot. (NASA/Ames)

(b) Taken by Voyager 1 on February 1, 1979, at a distance of 20,000,000 miles. This picture demonstrates the superiority of a television system over a simple spin-scan imaging system. Note how the pattern of swirling clouds north of the Red Spot has changed while to the southwest the dark area has changed to a light area, a tongue from which licks at the spot itself.

(JPL)

reported that Voyager's ultraviolet spectrometer had detected a doughnut-shaped ring, approximately 90,000 miles in diameter, consisting of ionized atoms moving around Jupiter in Io's orbit (figure 5-4). The ions were later found to be sulfur. Somehow these sulfur atoms had to be blasted into space from the surface of Io, a process at that time still unexplained.

Finding the mechanism was important, because sulfur ions appear to be spread throughout the Jovian magnetosphere. One theory was that the impingement of charged particles on the surface of Io might keep the surface stirred up, thereby accounting for the lack of ancient features such as impact craters.

Surprises came in rapid succession as the Voyager darted across the Jovian system. At the polar regions of the giant planet intense auroral activity was detected in the hydrogen spectra obtained by the ultraviolet spectrometer of Dr. A. Lyle Broadfoot. Au-

roras had been expected, but not as widespread as those discovered. The polar aurora was so intense that it could be seen by the instrument on the dayside of the planet.

It was weeks later, however, before Voyager's cameras photographed the aurora as the spacecraft was flying away from Jupiter on the outbound leg of its trajectory. Looking back toward Jupiter the spacecraft's cameras saw most of the planet's night side and only a thin crescent of illuminated hemisphere. On the pictures of the crescent-shaped planet the

Figure 5-4: Voyager was directed to pass through the flux tube (the narrow hoop) connecting Io to Jupiter. Jupiter is encircled by a cloud of hydrogen in a torus shape, the thick torus, and a cloud of sodium, the thin torus, both of which were explored by the Voyager spacecraft.   (JPL)

aurora appeared in the dark north polar region (figure 5-5). It was an astounding 18,000 miles long. And the pictures also revealed great groups of lightning flashes on the dark hemisphere of the mighty planet, thereby confirming the violent electric storms that before the flight scientists had predicted as likely to be raging on Jupiter. Lightning is important because, as explained in an earlier chapter, lightning flashes in an atmosphere containing methane, ammonia, and water can produce complex organic molecules that are believed to be the precursors of living things.

What appears to be a permanent jet stream of frozen ammonia swirls around Jupiter at 350 mph. Pictures of Jupiter (figure 5-6) became spectacular pieces of surrealistic art as the swirling cloud features were revealed in all their intricate detail.

Figure 5-5: An aurora stretching more than 18,000 miles is shown as the bright arc on this photograph taken from the dark side of Jupiter March 5, 1979, from a distance of 320,000 miles. The picture also shows several electromagnetic storms, the superbolts of lightning, which appear as the bright dots on the dark hemisphere of the planet. (JPL)

Figure 5-6: This image was obtained on June 29, 1979, when Voyager 2 was 5.6 million miles from the planet. The broad dark band extending across the lower half of the picture is the equatorial region of Jupiter. The turbulent region in the lower right-hand corner lies to the west of the Great Red Spot. High velocity winds along the southern edge of the equatorial zone combine with eastern winds to produce short-lived swirling cloud formations. (JPL)

While the spacecraft performed faultlessly, a thunderstorm in Australia on March 3 caused anxious moments at the Jet Propulsion Laboratory's Space Flight Operations Facility from where the mission was controlled. The Canberra receiving station missed nearly 4 hours of signals from Voyager, Nevertheless, all the rest of the time of the encounter, pictures came back to Earth in profusion. Time lapse motion pictures continued to be assembled and they now showed the swirls within the Great Red Spot in more and more detail, and small white ovals with spirals within them whirling by the spot along long braided flows of deep umber color. Some smaller storms appear to swirl around the Great Red Spot in a clockwise direction, circling the spot in about six days. Near the Great Red Spot a large white oval exhibited long streamers spiraling inside it (figure 5-7).

In one picture sequence of another part of Jupiter, dark-rimmed spots rolled along the edge of a light zone. One spot overtook another and whirled around it before plunging on ahead. The presence of dark-haloed bright spots along the boundaries between dark belts and lighter zones baffled scientists and did not fit earlier theories for the dynamic circulation of the Jovian atmosphere. These spots are deformed by the oppositely flowing jet streams but they are not absorbed into them. Scientists likened the cloud features to features seen between liquids that will not mix; like oil colors floating on top of a bucket of water and stirred gently, as used in the "marbling" that was at one time popular for the end papers of books.

Mysterious detail was apparent on the satellites (figure 5-8). Europa, which is about the size of Earth's Moon, had indistinct boundaries between light and dark areas, and sharp lines extending hundreds of miles across the otherwise featureless surface. Gan-

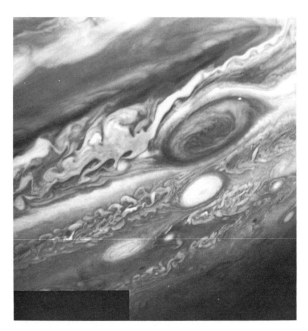

Figure 5-7: At a distance of 3.72 million miles from Jupiter, Voyager 2 obtained this remarkably clear picture of the Great Red Spot. A region of white clouds now extends from east of the Red Spot and around its northern boundary, preventing small cloud vortices from encircling the feature. The disturbed region west of the Red Spot has also changed since the time when Voyager 1 flew past Jupiter. It now displays more small-scale structure and cloud vortices that are being formed out of the wave structures. A white oval has moved beneath the spot; it is not the same oval that was beneath the spot when Voyager 1 flew by.   (JPL)

ymede showed linear features—great faults stretching thousands of miles across the planet-sized world, which is bigger than Mercury. An image of Io showed a series of circles like a dartboard, the outer circle being some 900 miles across. At the center of the concentric circles was a 100-mile-diameter dark spot. Scientists speculated—wrongly as was later shown—that the feature might represent a large impact basin. As more pictures were obtained, increasingly fascinating details were revealed on Io (figure 5-9). Its surface appeared scarred, reddish-gold in color, but very young looking, as though it had only recently been formed. Few craters were visible, but there appeared to be great areas of deposits, as though outflows of material from its interior had obscured all ancient features. Why Io should have such a sur-

Figure 5-8: These photos of the four Galilean satellites were taken by Voyager 1 during its approach to the planet in early March 1979. Io is top right, Europa top left, Ganymede bottom left, and Callisto bottom right. The satellites are reproduced to the same scale.   (JPL)

face was a major mystery. The solution to the puzzle would not, however, be forthcoming for several days and would be quite unexpected. Even the tiny innermost moon Amalthea had shown some detail. It was oddly shaped—twice as long as wide—and quite red in color.

There had been a complete change of viewpoint about the satellites. Before Voyager provided details of the surfaces of the Galilean satellites, scientists speculated, often somewhat wildly, about what the surfaces were like. After the Pioneers penetrated the Jovian system and obtained some tantalizing images, speculation increased. But when Voyager revealed the surfaces in intricate detail and showed them to be all quite different from each other, the task became one of trying to explain these differences.

In early March, a fascinating discovery was announced. It was late Friday evening, and past the really exciting phase of the close encounter. Scien-

Figure 5-9: Obtained March 2, 1979, at a distance of 1,600,000 miles, this picture shows the strange surface of Io, entirely unlike anything expected. White and orange patches appear as deposits, but no impact craters are visible. This is the first planetary body other than Earth to be found without a record of a period of intense cratering activity. (JPL)

Figure 5-10: The reason for Io's strangeness was discovered a short while later. Several days after the encounter, Linda A. Morabito, a JPL engineer, was examining this picture taken March 8, 1979, at a distance of 2.6 million miles, when she discovered two plumes which were later found to be volcanic eruptions, rising more than 150 miles above Io's surface. One plume extends from the limb of Io into space. The other is from the bright area into the dark hemisphere. (JPL)

tists were busily examining data. Many had left the Laboratory for the weekend to visit their homes or campuses in distant towns. Linda A. Morabito, a member of the project's optical navigation group, was processing pictures of Io taken against a star background so that the position of the spacecraft could be measured very accurately. The pictures showed most of Io in darkness with just a fine crescent illuminated by the Sun. When she had adjusted the gain on the reproduction of the pictures to make them suitable for the navigation measurements, Linda Morabito discovered a strange plume extending from the brightly illuminated part of Io and into the darkness (figure 5-10). She reported her discovery to Bradford Smith, head of the imaging team, but it was not until the following Monday that he could convene his fellow members for a conference to analyze the implications of her discovery. The scientists examined other pictures of Io closely, and some pictures taken earlier at close range were reprocessed

to search for similar plumelike features. The conclusion became inescapable. Io has active volcanoes. The obscuring matter was without doubt a volcanic plume extending 180 miles high above the satellite's surface. And there was more than one such volcano. The enhanced pictures showed cloudlike puffy features spread around the satellite. Io was a world in volcanic eruption, more active volcanically than Earth. Many strange features of Io could now be explained; the young-looking surface, volcanic mountains, lava flows, lack of ancient terrain and the sulfur in space along Io's orbit.

The discovery of Io's volcanoes was one of the biggest surprises of the Voyager mission, for before the discovery of the volcanoes Io had been expected to be very like Earth's Moon, not only in size but also in surface features. Its density was about the same, and the expectation was that its surface would be cratered and Moonlike, but perhaps with salt flats left by evaporation of saline solutions rising from its interior. Instead, the surface was being cooked, steamed, and fumed by at least six very active volcanoes. Three of these volcanoes were erupting at the same time, and all six erupted several times during the Voyager flyby. These active volcanoes could be correlated with vents on the surface. But undoubtedly, scientists reasoned, there are others which were inactive at the time of the passage of Voyager 1.

The plumes from the active volcanoes extended 200 miles into space, one to 300 miles. Data from the ultraviolet spectrometer revealed gases above the visible plumes to a height of more than 400 miles. To reach such high altitudes the material from the volcanoes had to be ejected almost explosively. The calderas themselves were shown to be extensive. They appear black on the reddish surface. One is located in a 1000-mile wide lava flow. Inactive calderas, as measured by the infrared instrument, are 500° F warmer than surrounding territory, implying that they contain hot lava lakes even though not erupting at the time of Voyager 1's flyby.

Now the question was the source of the heat driving the volcanoes. Some months earlier Stanton J. Peale, Patrick Cassen and Ray. T. Reynolds had submitted to the magazine *Science* an article suggesting that the interior of Io would be hot enough to be molten because of tidal forces from Jupiter, the nearby large satellite Europa, and the more distant Ganymede and Callisto. But Io is very dry and has no water, so the big question was how ash and volcanic processes could be generated in the material of Io. Another question was how long Io could continue spewing out material at the rate seen by Voyager without it becoming a hollow moon. One theory suggested that most of the material is redeposited on the surface and recycled, with only a very small part escaping into space to form the torus of ionized gas.

From its relatively high density. Io was thought to consist mainly of silicates, but a spectral analysis of the plumes was needed before scientists could decide what materials were being ejected by the volcanoes.

The discovery that the active volcanoes were spewing sulfur from the interior of Io answered some questions. Hot sulfur-rich magma also oozes through vents to produce the lavalike flows revealed in the pictures of Io's surface. The next question was whether the sulfur comes from deep within the interior of Io or from the bottom of a shallow crust, and what might be the nature of the driving gas or gases. Some of these questions would not be resolved until several months later when Voyager 2 could gather more information about this puzzling satellite.

By March 7 two other Galilean satellites had been revealed in detail, and they were found to be heavily cratered worlds.

Voyager passed within 71,000 miles of Ganymede on March 6 and within 78,000 miles of Callisto (figure 5-11) on March 7. Dr. L. Soderbloom of the imaging team explained at a press conference that Callisto was saturated with shoulder-to-shoulder craters over much of its surface, with the heaviest cratering around what appeared to be a large impact basin

Figure 5-11: Callisto looked very Moonlike when Voyager 1 discovered a great impact basin surrounded by concentric rings. But all the surface features of violent impact cratering were softened, as would be expected of a world of ice. (JPL)

similar to the lunar maria—the dark plains such as Mare Imbrium. However, the crater rims were not as sharp as might have been expected, certainly not as sharp as on the craters of the terrestrial planets and the Moon. While Callisto is most probably ice covered, at the very low temperature of the satellite ice behaves like steel. Callisto is not a very dense satellite, so scientists theorized that its crust is not thick nor strong enough to support high relief features such as mountains, crater rims, and scarps. An earlier suggestion was that the satellite is mainly water with a core of mud and a crust of ice. On such a surface the rims of craters would tend to flatten, as would all relief. Concentric rings around the large "impact basin" on Callisto also suggest a crust of very thin cold ice over a sea of warmer ice. They are reminiscent of ripples frozen in place on a pond.

Ganymede has fewer craters than Callisto. But one crater of several hundred miles diameter is the center for spreading rays of what appear to be debris. There are also bright spots, which might be interpreted as fresh ice, and darker shadings. Ganymede also has a complex system of sinuous ridges and grooves (figure 5-12) over the entire globe. One suggestion is that they represent deformations in the ice crust of this world.

There was another important and somewhat unexpected discovery during this same period. Voyager detected a stream of material inside the orbit of Amalthea, the closest known Moon to Jupiter. Mission controllers had programmed an 11.2-minute-long exposure on the camera when Voyager passed through the equatorial plane of Jupiter to look for dust or small satellites in this region. To the amazement of the scientists the picture showed a streaky image that could only be explained as a ring. The ring was discovered on March 4, but the results were so unexpected that the data were checked and rechecked before the discovery was finally announced on March 7. The ring is 34,000 miles above the upper cloud deck and well inside the orbit of Amalthea. The material of the ring is dark and appeared to consist of particles the size of boulders. It is less than 18 miles thick and about 5,000 miles wide. Scientists speculated that the ring originated from an asteroid

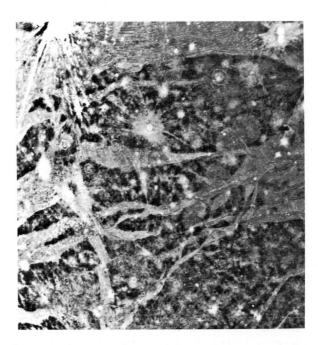

Figure 5-12: Ganymede was also surprising. Its surface displays many impact craters, numbers of which have bright ray systems. Bright bands traverse the surface and these contain intricate patterns of alternating bright and dark lines. This picture was obtained March 5, 1979, at a distance of 158,400 miles. (JPL)

that had strayed into the Jovian system and was broken up as it neared Jupiter.

By May, with the scientists still staggering under the load of new data about Jupiter and its satellites, Voyager 2 was activated and began its observatory phase of operations some 34 million miles from Jupiter. Approaching on a different sunline than did Voyager 1, the second spacecraft obtained a new viewpoint to obtain many pictures of the planet. Meanwhile Voyager 1 had been placed in the cruise mode on its way to an encounter with Saturn in 1980.

By June 1979, Voyager 2's pictures of Jupiter's clouds were becoming as detailed as those obtained by Voyager 1 (figure 5-13). But there were already striking differences. Important changes could be seen in the neighborhood of the Great Red Spot (figure 5-14); a white oval had drifted a considerable distance to the east of the Red Spot. A bright tongue of clouds ex-

Figure 5-13: Voyager 2's cloud pictures were equal in beauty to those of the first spacecraft. But features contrasted markedly with those observed by Voyager 1 (small photo). A bright tongue extending upward from the Great Red Spot is interacting with a thin bright cloud above it that has traveled twice around Jupiter in four months. A dark brown spot has developed (left photo) similar to the northern red spot observed by Pioneer. This was not present at the time of Voyager 1's flyby.
(JPL)

tending upward from the Red Spot was seen interacting with thin bright clouds to the north, a cloud system which in four months had travelled twice around Jupiter relative to other features. Turbulent waveforms to the west of the Great Red Spot which had been observed since 1945 now appeared to be breaking up. Also a dark spot had developed along the northern edge of the dark equatorial region. This

spot was similar to one seen by Pioneer in December 1973.

The spacecraft was still plagued by a few technical problems; a sticking filter wheel on the photopolarimeter, the inability of the receiver to lock on to a signal whose frequency is changing, and degradation of bonding material in the motor of the infrared instrument. But controllers had developed ways to work around these difficulties, at least in part.

Thirty-four days before its closest approach to Jupiter, Voyager 2 operated smoothly. But heating in one part of the spacecraft's main compartment still caused trouble with the remaining receiver on which a tracking loop capacitor had failed. The primary receiver had failed about the same time as the ca-

Figure 5-14: This pair of images shows two of the long-lived white oval clouds which have been in the Jovian southern hemisphere for nearly 40 years. The upper picture shows a cloud west of the Great Red Spot, and the lower picture is the cloud east of it. There is a third oval which was directly south of the Red Spot in 1979. These oval clouds display very similar internal structures. To the east of each of them, recirculating currents are clearly visible. In the lower picture there is a similar structure west of the white oval.   (JPL)

pacitor in April 1978. The effect was to limit the way in which the spacecraft could be commanded. Commands had to be sent to the spacecraft when the internal heating was at a minimum, which occurred when the spacecraft was not being oriented for spe-

cial maneuvers or when the power consumption of the spacecraft was not changing.

Commands for a backup mission had also been loaded into the computer memory so that the spacecraft would be able to continue to Saturn and make science observations there even if the receiver should fail completely.

In the period May 27 to 29, fifty hours of almost continuous picture-taking had assembled a series of whole disk images of Jupiter that would provide a time-lapse motion picture of five revolutions of the giant planet.

The spacecraft was aimed to fly past Jupiter at 60,000 mph at 7:29 P.M. EDT on July 9 with a closest approach of 444,000 miles above the cloud tops of the southern hemisphere. Voyager 2 would encounter Callisto (136,500 miles at closest approach) at 9:11 A.M. EDT on July 8, Ganymede (125,000 miles), Europa (34,000 miles), and Amalthea (347,100 miles) on July 9 at 3:06 A.M., 2:43 P.M., and 4:53 P.M. respectively, before the closest approach to Jupiter (figure 5-15). Table 5-1 compares the closest approaches made by the two Voyagers and the two Pioneers. Because Voyager 2 followed a trajectory past Jupiter that would allow it ultimately to fly to Uranus, it could not also fly close to Io. Io would be imaged from a distance after the encounter with Jupiter. There would be ten hours of observations to watch the volcanoes with the purpose of obtaining a time-lapse history of their eruptions. In addition, the spacecraft would attempt to photograph the thin Jovian ring again at two opportunities—23 hours before closest approach, and 39 hours afterward.

Table 5-1   Summary of Spacecraft Closest Approaches in Miles

	Pioneer 1	Pioneer 2	Voyager 1	Voyager 2
Jupiter	81,000	26,725	174,000	404,000
Amalthea	100,000	78,920	259,173	347,100
Io	221,300	164,981	12,738	702,056
Europa	372,500	364,760	454,962	127,900
Ganymede	496,400	430,130	71,209	38,700
Callisto	1,125,000	488,420	78,355	136,500

Voyager 2

	DATE	(ERT) Earth Received Time	KM	MILES	
CALLISTO	JULY 8	6:13 AM PDT	212,466	132,020	SURFACE
GANYMEDE	JULY 9	1:06 AM PDT	59,657	37,069	SURFACE
EUROPA	JULY 9	11:43 AM PDT	204,283	126,935	SURFACE
AMALTHEA	JULY 9	1:53 PM PDT	558,565	347,076	SURFACE
JUPITER	JULY 9	4:21 PM PDT	650,389	404,100	CLOUD TOPS
IO	JULY 9	5:09 PM PDT	1,128,030	700,925	SURFACE

NOTE: DISTANCES INDICATED ARE FROM SURFACE OF PLANET AND SATELLITES.

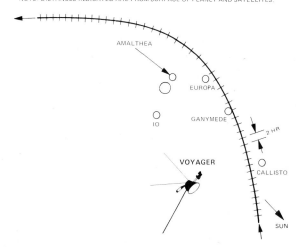

Figure 5-15: Voyager 2 made close approaches to several of the Galilean satellites and filled in details of the hemispheres not seen well during the encounter of Voyager 1 (see next chapter).   (JPL)

On June 27 the flight path of Voyager 2 was adjusted slightly. The spacecraft then crossed the bow shock on July 2 at a distance of 4.4. million miles from Jupiter. In the next three days the solar wind pressure ebbed and flowed so that at least eleven more crossings were recorded before the spacecraft was truly within the magnetosphere.

Taking advantage of the discoveries made by Voyager 1, the science sequences for Voyager 2 were modified. Scientists now had a clearer picture of what to look for within the Jovian system. They had discovered that Jupiter and its satellites do not resemble planets or satellites nearer the Sun, nor one another. The many unexpected findings were attributed to real differences between the outer and inner Solar

System and to lack of prior knowledge about the outer planets, caused largely by the enormous distances to them making observation difficult from Earth. The scientists wanted a clearer picture of Europa to resolve the strange linear markings seen indistinctly on the satellite by Voyager 1. Scientists had speculated that these might be cracks in an ice sheath covering a rocky surface. There was also some debate as to whether or not vulcanism might have been induced on Europa as on Io. Three types of volcanoes had been identified on Io: those with lava flows, those blowing plumes high above the surface, and broad calderas which changed color in the few hours of observations made by Voyager 1.

The imaging program was also directed toward obtaining views of parts of the surfaces of Callisto and Ganymede not seen during the Voyager 1 flyby. The spacecraft would also seek to find out more about the Io torus of ionized sulfur and oxygen atoms in which some of the particles have tremendous kinetic energies and give rise to a plasma with an intensely high temperature. The spacecraft would also attempt to gather more information about the auroras and lightning flashes on the night side of Jupiter.

July 8: early in the morning, Voyager 2 photographed the ring of Jupiter, one of the major scientific objectives of the second encounter. Pictures were obtained from slightly above and slightly below the ring to try to obtain information about its form.

Later that day pictures arrived at Earth showing tremendous detail of Ganymede's great crustal cracking and slippage. The following day equally clear pictures began to arrive at Earth showing close-ups of the surface of Europa. And in the early afternoon July 9 Voyager started a photo sequence imaging the tiny inner moon Amalthea as the spacecraft zoomed within 347,076 miles of its rugged red surface. At this time L. Soderbloom of the imaging team commented about the astounding discoveries concerning the Jovian satellites so far.

"Some months ago," he said, "before the Voyager 1 encounter, we thought we had some idea of what planets were like, at least the planets of the inner Solar System: Mars, Mercury, Moon, and Earth. We've discovered many times over in the last couple of months how narrow our vision really was. Included in the Galilean collection of satellites are the oldest, the youngest, the darkest, the brightest, the reddest, the whitest, the most active, and the least active. Today we found the flattest.''

Europa was shown by the pictures to be almost billiard-ball smooth with its surface fractured by many thousands of dark and light lines like a cracked eggshell. Absence of craters and topographical relief suggested that Europa might be a relatively young satellite. Some scientists speculated about Europa having been liquid slush at one time and then, as the satellite cooled, the surface froze into a crust of ice. Later freezing pushed up the surface and produced the many cracks in it.

Io and Europa appeared to be satellites active because of tidal forces and internal heating, but Europa much less so than Io. Ganymede and Callisto, however, seem more ancient worlds with many craters and impact features. Callisto is saturated with craters whereas Ganymede has many areas with only few craters but a variety of fault lines. Neither world has high mountains or great differences in topography. Concentric ring formations on ancient Callisto point to impacts of asteroid-sized bodies in the distant past, impacts that spread shock waves thousands of miles over the surface. On Ganymede there is evidence of an even greater impact the result of which was to wipe out vast areas of ancient cratered terrain.

July 10: as Voyager 2 flew behind Jupiter, Io emerged from occultation by the great bulk of the planet. With the Sun shining from behind the satellite and Jupiter, both worlds appeared as illuminated crescents to the eyes of Voyager's television cameras. Voyager completed its 150-photo, 10-hour volcano watch of the sulfurous moon. One large volcano (plume 1) seen erupting at the passage of Voyager 1 was now quiet. But others were still active. No new volcanoes were seen. Also the footprint of plume 1 on the surface had disappeared as though covered with debris. The

extensive surface feature had completely vanished during the few months between the two encounters.

Scientists offered several theories for vulcanism on Io. In one the satellite is visualized as having a surface crust of solid sulfur and sulfur dioxide beneath which is a slurry of solid and liquid sulfur dioxide. Beneath that is a worldwide ocean of liquid sulfur surrounding an interior or silicate rocks. In this ocean is the energy force for the volcanoes, analogous to the molten rocks of Earth. Magma rising through the rocks from the heated interior of Io forces molten sulfur to the surface, driven through volcanic vents by sulfur dioxide gas. Another theory was that the silicate magma is covered with a crust of solid silicate with only small seas of molten sulfur and a relatively thin crust of solid sulfur and sulfur dioxide. In this theory the vulcanism is silicate-based rather than sulfur-based.

Sulfur and frozen sulfur dioxide can provide all the rich coloring seen on Io. There are many crystalline forms of sulfur all of different color, and sulfur dioxide snow is white. Later inspection of the Io pictures showed the satellite vents glowing blue material— probably sulfur dioxide condensing into snow— from its south polar region along faults or fractures in the crust. Io's vulcanism appeared to be analogous to terrestrial vulcanism but operating in a very cold environment, hardly much above the boiling point of water on Earth.

A third theory visualizes the thin crust of Io as consisting of sulfur dioxide mixed with sulfur, topping a layer of solid porous sulfur mixed with liquid sulfur dioxide. When the ocean of liquid sulfur periodically breaks through into the crust and comes into contact with the cold liquid sulfur dioxide the latter flashes into vapor and releases energy sufficient to cause an eruption. This theory also explains the sheer cliffs on Io that are hundreds of miles long. Geyserlike eruptions of sulfur dioxide probably erode these cliffs continually and this gas condenses to form the deposits of white snow seen at their bases.

As the spacecraft pulled away from Jupiter the instruments and cameras gathered additional information about the auroras and lightning bolts first detected by Voyager 1. Auroral emissions in the polar regions were observed in ultraviolet and visual light. These auroras are generated at three altitudes about 435, 870, and 1,430 miles above the tops of the clouds. They extend from polar regions to latitude 60 degrees. Lightning was observed planetwide, appearing in clusters. Individual flashes are extremely powerful, equivalent to superbolts of terrestrial lightning.

July 10: a second look at Jupiter's ring. Clear, sharp pictures of the ring were sent to Earth, and Bradford Smith said that the ring appeared larger than expected and of more complicated form. More ring pictures were taken the following day with the spacecraft looking back on the dark side of Jupiter. The ring is extremely thin—perhaps only half a mile thick—and as narrow as 4,000 miles. It is about 33,000 miles above the cloud tops of Jupiter and appears like a fine bracelet around the planet. But the material does not appear to be consistent all around the ring. There appears to be some material inside the ring, as though clouds of ring particles are being pulled from it and falling to the top of Jupiter's atmosphere where they would produce a spectacular and continuing display of meteors.

At the time it was speculated that the ring consisted of particles that should have formed into a satellite of Jupiter but were prevented from doing so because they were too close to the giant. Another theory was that the ring particles originate from the volcanoes of Io. This is not believed likely, however, because the outer edge of the ring is sharply defined; there seems no trace of any material coming into the ring. Later a tiny satellite of Jupiter was discovered (by G. Edward Danielson and David Jewitt) hurtling around Jupiter at 67,000 mph just outside the ring. The satellite, Jupiter's 14th, is a dark object only about 20 miles in diameter. A conclusion was that the ring originated either from this satellite or from another like it that had broken apart. Both could be captured asteroids.

As each Voyager approached Jupiter it made a long sequence of color images taken every 72 degrees of longitude, which is one-fifth of the rotation of the planet. Within 12 days of the spacecraft's closest approach Jupiter appeared too large from the spacecraft for mosaics of the whole disc to be obtained. Instead the cameras zoomed in on selected targets of greatest interest to the scientists (figure 5-16). After closest approach further sequences of images were obtained as the spacecraft sped away from Jupiter. These images showed the planet in a crescent phase and the pictures covered both the illuminated crescent and the dark gibbous part of the planet's disk.

From these pictures, numbering many thousands, scientists were able to resolve many questions about the motions of the Jovian clouds. It was concluded that, because there are uniform wind speeds for cloud features with widely different sizes, the cloud motions must be attributed to motions of masses of atmosphere rather than wave motions within the atmosphere. The scientists saw white clouds rapidly forming and spreading, these features being attributed to disturbances that caused convection in the atmosphere.

A pattern of east-west winds was found in the polar regions where earlier observations by Pioneer had suggested that the main activity was cellular, caused by convective upwelling and downwelling of masses of atmosphere.

Within the Great Red Spot the atmosphere rotates anticyclonically in a period of about six days. And interactions between small spots and the Great Red Spot were observed. The atmosphere above the Great Red Spot was shown to be substantially colder than that in surrounding regions.

The full disk of Jupiter, spread into a cylindrical projection, differed markedly between the two Voyager encounters, as illustrated in figure 5-17.

The highest wind velocities were measured at the boundaries between belts and zones. Wind velocities were measured by tracking small features. Eastward

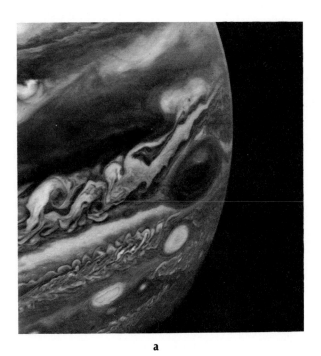

a

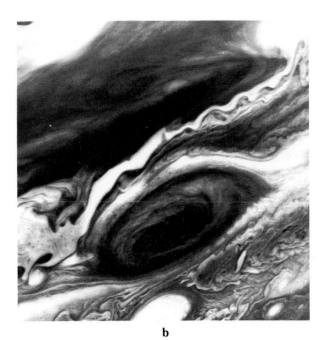

b

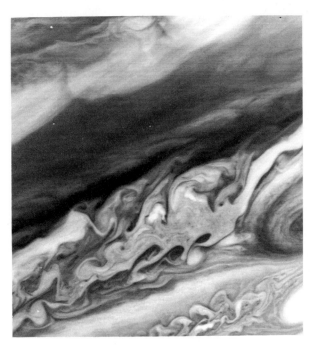

c

Figure 5-16: During the flyby of Voyager 2 scientists directed the cameras to obtain more details of targets of interest first detected by Voyager 1. These many pictures, of which these are examples, revealed a wealth of new detail about the cloud systems of the giant planet.

(a) North of the Great Red Spot lies a curious darker section of the South Equatorial Belt. A bright eruption of material passing from the belt into the diffuse equatorial clouds north of it has been observed on all occasions when this feature passes north of the Red Spot. The remnants of such an eruption are seen in this picture.   (JPL)

(b) This picture shows the Great Red Spot and the Southern Equatorial Belt extending into the equatorial regions. The clouds in the equatorial zone are more diffuse and do not display the structures seen in other locations. Considerable structure within the Red Spot is visible in this picture.   (JPL)

(c) This picture provides details of the intricate patterns west of the Great Red Spot. These features move along the edge of the equatorial zone, but the remainder of the region is characterized by diffuse clouds. The region west of the spot is seen as a disturbed wavelike pattern. Similar wave patterns are seen west of the white oval at the bottom right of the picture. (JPL)

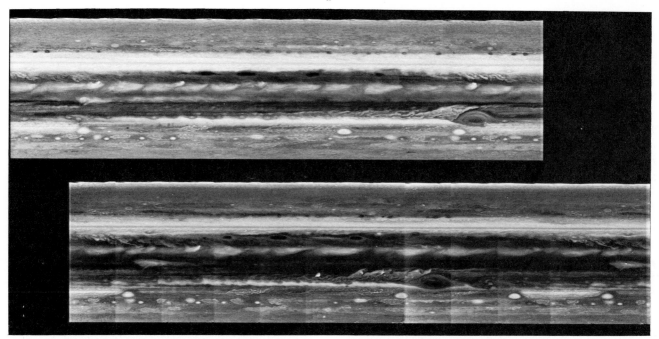

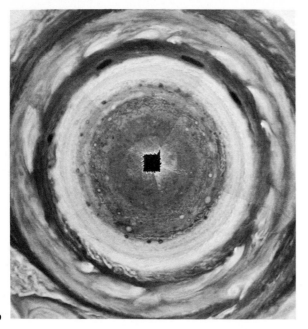

b

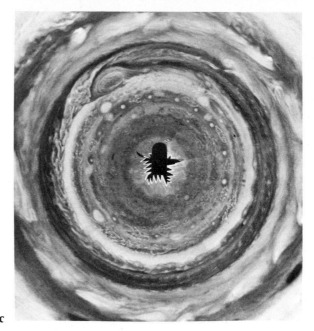

c

Figure 5-17: Projections were made from the Voyager pictures to provide details of the Jovian weather systems.

(a) Cylindrical projections compare Voyager 1 and Voyager 2 images. The pictures are aligned so that the longitude scales are correct. It can be seen that the Great Red Spot has moved westward and the white ovals eastward during this period. Regular plume patterns are equidistant around the northern edge of the equator, and a train of small spots, at approximately 80 degrees south latitude, has moved eastward.   (JPL)

(b) Another projection constructed from Voyager 1 images shows the giant planet viewed from above its north pole. The belt/zone structure extends to at least 50 degrees north latitude. At the northern edge of the equatorial region plumes are positioned equidistantly around the planet. The north temperate region's high speed jet appears as a thin line.   (JPL)

(c) The south polar projection shows the Great Red Spot and the white ovals. Again the belt/zone features become less distinct at high latitudes but discrete cloud features still appear to be positioned zonally.   (JPL)

currents were located at latitudes 47, 41, 35, 23, and 8 degrees north, and at 8, 23, and 32 degrees south. Westward winds were located at 50, 44, 38, 30, and 18 degrees north, and at 16, 28, and 35 degrees south. An eastward wind blows along the equator at 200 mph. Velocities as high as 330 mph were measured at subtropical latitudes.

For the first time intricate detail was observed within the bright ovals (figure 5-18) showing that they also are anticyclonic like the Great Red Spot.

The Pioneer pictures were interpreted as showing that the white plumes were trailing westward behind small bright origins near the equator. High resolution pictures from the Voyagers confirmed that this is so.

Figure 5-18: This picture shows a region of the southern hemisphere extending from the Great Red Spot to the south pole of Jupiter. The white oval is seen beneath the Red Spot, and several smaller spots are further to the south. Some of these organized cloud spots have similar morphologies, such as anticyclonic rotations and cyclonic regions to their west. The presence of the white oval causes the streamlines of the flow to bunch up between it and the Great Red Spot.   (JPL)

Rapidly ascending atmosphere appears to produce bright material which then disperses to the west. These features have small individual puffy heads like cumulus clouds. It may be that a wave pattern flows around Jupiter and triggers formation of these white features from which the plumes are generated.

Voyager confirmed the Pioneer and ground-based interpretations of the dark belts, the brownish spots, and the blue grey regions near the equator as being places where the atmosphere is sinking. These regions are warmer than their surroundings which indicates that they are lower in the atmosphere. In one image the high white clouds can be seen spreading over a brown spot (figure 5-19). Their reddish color must result from a different process from that which produces the red color of the Great Red Spot.

A fascinating discovery by Voyager was that the high latitude cellular structure discovered by Pioneer is in fact structured internally with spiral forms similar to those within the large ovals and the Great Red Spot. Also the Voyagers discovered that the east-west velocities extend into high latitude regions.

The time-lapse pictures revealed how small spots interact with the Great Red Spot. They generally approach it from the east, flow around the equatorial side of the spot, and then either continue around the planet leaving the Great Spot behind or else become absorbed into it. When smaller spots of similar size come together because one is overtaking the other, they often merge into one spot and emit a streamer toward the equator and the west.

Andrew Ingersoll commented on a regular pattern of motion discovered in the time-lapse cloud pictures of Jupiter itself. The big question facing scientists is why the banded nature of the Jovian clouds and features lasts for hundreds of years. How can such turbulent features seen in the Voyager pictures be so orderly? "At first Voyager seemed to do nothing but emphasize the chaos, not the order," he said. "This is the order that we did find in all the chaos. It is a regular pattern, a very persistent, almost stationary pattern of eastward and westward velocities in spite of the changing colors and visual appearance

a

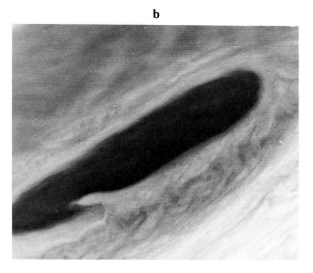

b

Figure 5-19: Evidence was obtained that the light clouds are higher than the dark clouds.

(a) The dark belt across the upper part of this picture is the North Equatorial Belt. The dark elongated oval is one of the largest of the long-lived features along the northern edge of this belt. (JPL)

(b) A high, white cloud is seen moving over the darker cloud of the oval, providing an indication of the structure of the cloud layers. Thin white clouds are also seen within the dark oval. (JPL)

on the surface." Every pattern that has been recorded by Earth-based observations over the last 75 years was seen repeated in the 10-hour sequence of pictures taken by Voyager.

"What exactly this order means is less clear, but most of us are beginning to pay much more attention to the interior of Jupiter, or at least the layers of Jupiter below the visible clouds. And the turbulence that we see in the visible clouds is kind of a minor side show."

The pattern appears to be one of small-scale chaos and large-scale order. Atmospheric cloud patterns seem to be tied to flow patterns deeper in the interior.

The Great Red Spot had changed since Voyager 1's flyby. It had become noticeably lighter in color. The flow patterns around the spot had also changed, with a reversal of flow to the east.

Ed Stone summarized some of the important science results from the flybys: "Voyager managed to make dynamic measurements of the weather systems on Jupiter. It showed, for example, that the Great Red Spot is an anticyclonic vortex. We can actually now start modeling the upwelling of the material in the formation of the Red Spot. It has also been possible to show that in addition to the east-west alternating jets not all winds are purely east or purely west; they tend to have random variation, eddies. The question is, are the eddies being generated by the jets or are the eddies generating the jets? It has been possible to show by statistical analysis that the eddies are feeding energy into the jets. The basic idea is that you have an atmospheric layer and . . . because of convection effects from beneath, due to the internal heat source, the layer is occasionally bumped. When you bump something you squeeze it and that generates a small eddy, and energy which is caused by this bumping eventually is transferred into the jet velocity. The amount of energy is 3 or 4 watts per square meter being dumped into the atmosphere by this eddy effect.

"It has been possible to get a handle on the source of the energy and the nature of the dynamics of the

Jovian atmosphere for the first time. It's clear that the auroral activity that was discovered in the polar regions is very important to an understanding of the chemistry of the upper atmosphere of Jupiter. There is such a tremendous influx of energetic particle energy that it is driving a lot of chemical reactions, and it may, in fact, produce the dark haze which was discovered by Voyager in the polar regions.

"The ring around Jupiter is . . . one more example of a ring system that we hope will contribute to our eventual understanding of ring systems. It may turn out that these ring systems are much more similar than we thought. The Jupiter ring is clearly a transient ring. I think that it is transient in the sense that the material can be seen spiralling from the ring to the top of the atmosphere. So material is being continually lost to the atmosphere, but the fact that it is there means that it is also continually being replenished. So it is in that sense a dynamic ring. That may not be the case with Saturn's rings. There are a number of satellites just outside the ring that may be supplying the material, and there might also be large bodies within the ring supplying the material. If that is so, eventually the ring will disappear. But there is no evidence that the material of the rings of Saturn is being lost. It seems to be constrained within the ring system.

"The most spectacular discovery was the Io torus, which is being fed in some way by the volcanic activity on Io. This torus seems to be the missing link in our trying to understand a number of things about the Jovian magnetosphere. It supplies sulfur and oxygen plasma throughout the magnetosphere; we actually saw particles, moving at ten percent the speed of light, which had been accelerated within Jupiter's magnetosphere. The torus was glowing in ultraviolet; glowing sulfur and oxygen at a temperature of over 100,000 K. There are really two toruses. The hot big one, and inside that a smaller cold torus which can be seen from Earth, cold enough to emit in the visual [visible light], but it is the same material. The torus presumably is somehow responsible for the intense auroral activity in the Jovian atmosphere.

"The torus is responsible for the fact that we missed the Io flux tube. In a sense the torus is so dense that the electrical signal in a plasma is merely a pluck on a magnetic field line. The disturbance on a magnetic field line generated by Io travels very slowly in a dense plasma. The net result is that when it escaped from the torus it zipped down to Jupiter and came back and then slowed up and, when it finally got back to Io, Io had moved. What had happened was that the magnetosphere had rotated so that the current system did not close back on Io. And as we had targeted to go beneath Io the current system had already moved. By the time it had gotten so far from Io it had moved sideways. We detected the flux tube remotely because the current system generated a magnetic field which could be sensed from a distance. So we know that there were about 3 million amperes flowing in the flux tube."

While the new science results were stimulating many discussions, there had been some problems with the instruments of the spacecraft. The photopolarimeter had failed during the Jupiter encounter of Voyager 1. The cause of failure was not known. The polarimeter on Voyager 2 also failed, but only partially. Only four of the filter wheels could be commanded into operation. In the instrument there is an integrated circuit which counts alternate numbers instead of every number.

The infrared instrument had developed some problems during cruise but these had been overcome for the encounters. It operates at 200 K and measures the interference patterns of the infrared beam coming into the instrument. Any misalignment of the mirrors even a fraction of a wavelength would result in a problem. The beam splitters and some of the mirrors in that instrument were mounted with a silicone compound which at 200 K evidently crystalized, stiffened, and changed its bulk properties enough to shift the alignment and degrade performance. These were materials properties which no one knew about because nothing similar had previously been flown at 200 K for several years in space. The engineers developed techniques to heat it during cruise and keep it warm and then allowed it to cool off for the planetary observations.

Immediately after the closest approach the course change maneuver which had been stored in the spacecraft's memory was implemented. The maneuever was needed to direct the spacecraft on a path that would take it to Saturn and maintain propellant reserves as discussed in the previous chapter. Two hours after Voyager skimmed over Jupiter's cloud tops, four of the spacecraft's thrusters fired for 76 minutes. "By doing the maneuver during Jupiter flyby," explained deputy project manager Esker K. Davis, "twenty-one pounds of control fuel will be saved, which will make possible a journey to Uranus after Voyager 2 passes Saturn in August 1981."

The maneuver had, however, caused some extremely anxious moments because of some problems with the spacecraft's communications. Kohlhase pointed out: "A passenger aboard Voyager 1 in its passage through the radiation belts of Jupiter would have received a total of 400,000 rads, which is 100 times the lethal dosage. Even if shielded, the passenger would have received 50 times a lethal dosage." Although Voyager 2 flew farther out from Jupiter it still had to contend with intense radiation by terrestrial standards. The ailing command receiver, and the fact that there was no backup receiver, worried project management and scientists because radiation levels were high enough to cause problems in the faulty receiver.

But all went well and Voyager 2 followed Voyager 1 toward Saturn. After Voyager 1 encountered Saturn it would fly out of the Solar System, whereas Voyager 2, if all went well with the spacecraft, would have the capability of flying to Uranus and then on to Neptune.

6.

Close Looks at Satellites

ALTHOUGH there were many surprises about Jupiter during the Pioneer and Voyager missions, and undoubtedly there will be more when Project Galileo can orbit the giant planet and send a probe deep into its atmosphere, by far the most exciting and unexpected discoveries were made about the Galilean satellites. As a result of the Voyager mission to Jupiter the number of unique bodies in our Solar System on which we have detailed information was virtually doubled. Prior to the mission to the outer planets, the Earth-sized bodies—Mercury, Venus, Earth, Moon, and Mars—were shown, as the result of space missions, to be unique worlds, each quite different from the others, and each telling a unique section of the history of the formation and subsequent evolution of the Solar System. In the mission beyond Jupiter to Saturn another new class of satellite worlds, again with remarkably different characteristics among its members, was revealed.

Before Voyager the Galilean satellites had been objects of much speculation. When Earth's atmosphere is particularly steady the Galilean satellites can be seen as disks through a powerful telescope. But the satellites are less than one second of arc in diameter when viewed from Earth, so that it is virtually impossible to be sure of any detail upon their surfaces. Io was known to have dark polar areas from photographs obtained by equipment carried high in Earth's atmosphere by a Stratoscope balloon in 1973. And some astronomers made maps of the diffuse light and dark markings on the satellites. But little real progress could be made until a spacecraft could visit the giant planet. The glimpses of the satellites revealed by the spin-scan imaging systems carried by the two Pioneers, although indistinct and lacking detail, pointed toward subsequent discoveries.

In the 1920s careful observation of the satellites had shown that their brightnesses vary with their position in orbit, implying that there are areas of surface with different reflectivity. The cyclic nature of the changes also indicated that the satellites are tidally locked to Jupiter so that, like Earth's Moon, each rotates on its axis in the same period it takes to revolve around the parent planet. Continued observations showed also that the surfaces of the Galilean satellites differ greatly from each other. Io has the highest reflectivity and Callisto the lowest. Another peculiar fact is that the leading hemispheres of Io, Europa, and Ganymede are lighter than the trailing hemispheres, whereas the reverse is true for Callisto.

Temperatures on the surfaces of the satellites were measured in 1962, and only reached $-160°$ to $-207°$ F, even at noonday on each satellite. By measuring the changes in infrared radiation as the satellites enter and emerge from the shadow of Jupiter, astronomers obtained evidence of surfaces which appeared to be of fine rocky fragments or crystals of frost. Frost was also indicated by spectroscopy in the mid 1950s, and was confirmed in subsequent decades by more refined observing techniques, but strangely, not for Io or Callisto. These satellites seemed to have rocky surfaces. Io's surface characteristics could be accounted for by a surface covered with sulfur or salts. But the absence of ice on Callisto was mysterious because the outermost satellite was thought to be rich in water. Masses of the satellites were determined to within 10 percent and their diameters measured to within 5 percent by astronomers timing occultations of stars by the satellites. Their densities calculated from these measurements showed decreasing value with increasing distance from Jupiter. Io and Europa have densities similar to our moon, and Ganymede and Callisto have densities less than twice that of water. So Io and Europa appeared to be rocky bodies while Ganymede and Callisto appeared to consist mainly of water.

But there were strange anomalies. While Europa was thought to have lost most of its water, its surface appeared to be covered with frost. But Callisto, with the most water, did not display water frost on its surface.

In 1973 R.A. Brown detected a cloud of sodium gas extending far into space from Io into a toroidal shape along the orbit of the satellite. Later, potassium and sulfur were identified in the cloud. And Pioneer dis-

covered a hydrogen cloud too. No similar clouds connected with the other satellites were detected. The question of why Io of all the satellites should be so equipped remained unanswered until Voyager.

Sophisticated new techniques of observation from Earth, while not revealing surface details on the big satellites, provided information about their environment and their surface characteristics. The inner moons were thought to be rocky, the outer moons more like giant balls of icy water, with increasing amounts of water at increasing distances from Jupiter. Io was thought to be a rocky world covered with salt flats, Callisto a predominantly icy world (figure 6-1). Callisto's surface was shown to be water bonded to surface material and not free water ice. Very little water ice is present on the surface of Callisto despite the satellite consisting mostly of water. Also it was found that more than half the surface of Ganymede is covered with water ice, while the darker areas seemed more akin to the surface of Callisto. One explanation was a competition between dust falling from space onto their surfaces and water being brought from their interiors to recoat the surface. Infrared observations of Io from high altitude aircraft showed that its surface bore little resemblance to any other object in the Solar System. There was no evidence of water or ice. Speculation was that the surface might have been formed by salt deposits brought to the surface and left there when the water subsequently evaporated.

In 1975 the surface of Ganymede was probed by radar from Earth and found to be rougher than any of the inner planets. The surface was explained as a rocky or metallic material embedded in a matrix of ice. Such a surface would appear smooth optically but rough to radar because of the transparency of ice to radar frequencies.

Voyager painted an intriguing new picture of the Galilean satellites. While they did, indeed, appear to have increasing amounts of water at increasing distances from Jupiter, each satellite was revealed as a unique body, in much the same way that the inner planets are unique from each other. The inner two

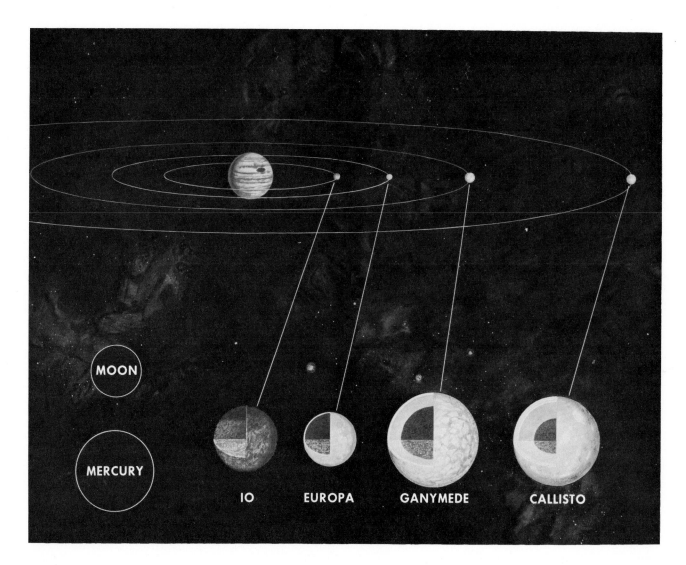

Figure 6-1: The Galilean satellites of Jupiter were thought to be quite different worlds even before the mission of Voyager. Because their densities vary greatly from the innermost to the outermost, Io was thought to be a predominently rocky world, and Europa the same, with possibly a crust of ice on Europa. Ganymede and Callisto were thought of as ice worlds, with small rocky cores surrounded by deep oceans of water and capped with a crust of ice. (JPL)

satellites, Io and Europa, possess entirely different surfaces, and neither one corresponds with the two outer satellites, Ganymede and Callisto, which themselves possess quite different surfaces from each other.

Of the rest of Jupiter's sixteen satellites (three of which were discovered by examination of images returned by Voyager) only Amalthea was inspected by the spacecraft. The other satellites are too small and were too distant for imaging them effectively. All these satellites are very small bodies and many are believed to be captured asteroids. The satellites are in two dynamical groups (figure 6-2). One group located at a distance of about 164 radii of Jupiter

JUPITER MOONS

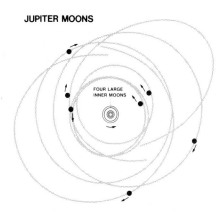

Figure 6-2: The many other small satellites of Jupiter were not approached by the Voyagers. The satellites are in two major groups; one group has members that orbit in the same direction as Jupiter rotates, the outer group orbits retrograde. Compared with the Galilean satellites these two groups are at great distances from Jupiter, as shown in this drawing. (NASA/Ames)

from the planet consists of small satellites moving in elliptical orbits with inclinations of about 27 degrees. The second group, located at a distance of about 322 radii of Jupiter has satellites moving in

retrograde orbits. It has been suggested that Jupiter captures comets and their nuclei might remain in orbit about Jupiter for 100 years or so. Such comets might account for very small satellites that have been observed and then lost. Observations of these small satellites from spacecraft will have to wait for future missions to the giant planet. The known satellites of Jupiter are listed in table 6-1.

Amalthea

Jupiter's innermost satellite, Amalthea was discovered in 1892, the first satellite of Jupiter to be discovered after the discovery of the Galilean satellites in 1610. Being so small and so close to Jupiter—the radius of Amalthea's orbit is only 112,473 miles—Amalthea is difficult to observe from Earth. But astronomers knew that it is colored red and does not reflect light well. Revealed by Voyager's camera (figure 6-3) Amalthea's surface is dark and red and quite different from the surfaces of the Galilean satellites.

Table 6-1   The Satellites of Jupiter

Number	Name		Mean Orbital Radius		Period	Discovery
	current	old	miles	Rj	days	year
16	1979J3	—	79,370	1.79	0.29	1980
14	Adrastea	1979J1	83,268	1.89	0.30	1979
5	Amalthea	Amalthea	112,473	2.55	0.49	1892
15	1979J2	—	137,950	3.13	0.67	1980
1	Io	Io	262,230	5.95	1.77	1610
2	Europa	Europa	416,960	9.47	3.55	1610
3	Ganymede	Ganymede	664,898	15.10	7.15	1610
4	Callisto	Callisto	1,168,232	26.60	16.70	1610
13	Leda	—	6.903,754	156.00	240.00	1974
6	Himalia	Hestia	7,127,458	161.00	250.57	1904
10	Lysithia	Demeter	7,276,594	164.00	263.55	1938
7	Elara	Hera	7,295,236	165.00	259.65	1904
12	Ananke	Adrastea	12,862,980	291.00	617.00	1951
11	Carme	Pan	13,888,290	314.00	692.00	1938
8	Pasiphae	Poseidon	14,478,620	327.00	739.00	1908
9	Sinope	Hades	14,727,180	333.00	758.00	1914

NOTES: Orbits of the outer satellites vary considerably in relatively short times.

Names of satellites (other than Galilean) were changed to those recommended by an International Astronomical Union Task Group for outer Solar System nomenclature, whose report was published in 1976.

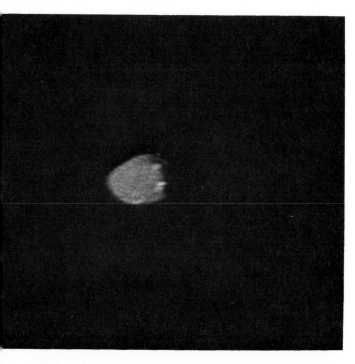

Figure 6-3: Voyager did, however, look at one satellite other than the Galileans. This is the innermost satellite, Amalthea (innermost until Voyager discovered even closer small satellites). Tiny, red Amalthea whirls around the planet in 12 hours, only 1.55 Jupiter radii above the cloud tops. This view is from 255,000 miles on March 4, 1979. Amalthea's irregular shape probably results from a long history of impacts with other bodies in the Jovian system. The satellite keeps its long axis pointed toward Jupiter. (JPL)

Amalthea is a small, battered, elongated object about 165 miles long and 102 by 93 miles across, with its long axis pointing toward Jupiter throughout its 12-hour orbit. Scars on its surface include large craters, sharp ridges, and a general topography that implies a long history of cosmic bombardment. The largest crater, Pan, is about 56 miles in diameter. Several bright spots are visible in the images of this small satellite. Their composition is unknown, but they have a slightly greenish hue and seem to be located on slopes or ridges.

Amalthea's surface is quite red, redder than typical Trojan asteroids which travel around the Sun in Ju-

piter's orbit, but it is not as red as the surface of Io. Possibly the surface has been contaminated by the sulfur which Io's volcanoes contribute to the Jovian environment. The surface is also heavily bombarded by high-energy-charged particles of the Jovian magnetosphere which probably contribute to its color.

Io

Io proved to be the greatest surprise of all (figure 6-4). As recounted earlier, not only is it more active volcanically than any other body in the Solar System, but also it spreads sulfur throughout the Jovian system and interacts strongly with the magnetosphere and its energetic particles.

Io has a wide distribution of many different volcanic forms (figure 6-5): calderas, shields, rough-surfaced lobate flows, digitate flows, blankets of debris originating from eruptions, red and orange pyroclastic plains, and other deposits of pyroclastics. All these different forms indicate that sulfur-based vulcanism is very similar to silicate vulcanism (the activity on Earth) or that Io is undergoing silicate vulcanism which is masked by planetwide deposits of sulfur.

One of the calderas on Io has a wall that is about 1.2 miles high. The slope of the material making up the wall is about 45 degrees. Preliminary analysis indicates that this feature could not be composed entirely of sulfur because that material would not have the strength to form such a slope.

In the past Io has been observed to brighten considerably on emerging from Jupiter's shadow. This brightening may have been caused by sulfur-dioxide frost when Io's volcanoes are very active. Some scientists have speculated that Io might have been at a low level of volcanic activity during the flybys of the two Voyagers. There is some statistical evidence from posteclipse brightening that Io's volcanic activity cycles in a period of about 2 years; that is, if the brightening is caused by an increased amount of sulfur dioxide which arises only from volcanic activity.

a

b

c

Figure 6-4: Io was perhaps the most interesting of all the satellites because it proved to be currently more active volcanically than our own Earth.

(a) Io is shown here in front of Jupiter, showing how relatively small it is compared with the Great Red Spot and other Jovian features. Yet Io is as big as Earth's Moon.   (JPL)

(b) This picture taken at 2.9 million miles shows intriguing detail on Io. The strange feature on the lower left of the disk, which was originally thought to be a large impact basin, is, in fact, a great volcano, Pele.   (JPL)

(c) This full disk image was made from several frames taken by Voyager 1 on March 4, 1979. Many different types of features can be seen. The circular, doughnut-shaped feature at the center is an erupting volcano. Other similar features are seen elsewhere on the disk. White patches are probably deposits of sulfur-dioxide frost and snow. Dark areas are calderas, and there are many spreading lava flows.   (JPL)

Infrared observations of Io from Voyager revealed many thermal anomalies. The highest temperature, associated with the volcano Pele (figure 6-6), is far above the melting point of sulfur (234° F). This indicates that there must be vigorous convection and flowing of high-temperature molten sulfur through cracks in the crust. Many of the hot spots observed on Io were not associated with eruptive plumes, which implies that there are regions that may erupt

a

b

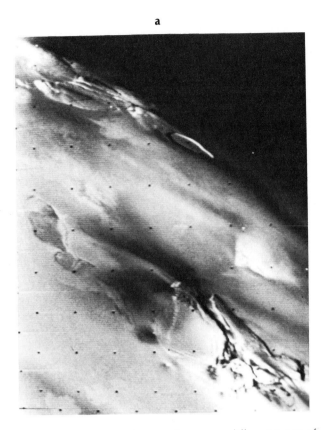

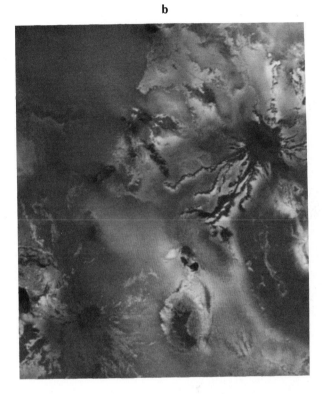

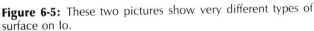

Figure 6-5: These two pictures show very different types of surface on Io.

(a) Taken on March 5, this picture is of an area around latitude 15 degrees south and longitude 244 degrees. Many depressions and elevations are apparent. The sunlight is coming from the left, so a depression has a bright right wall and a shadowed left wall. These depressions have complex shapes and do not resemble impact craters. Two are joined by a narrow trough. The elevations are irregular and are cut by linear and irregular troughs. The surface is smooth like a plain and not pockmarked with craters as are the plains of other worlds.   (JPL)

(b) Also taken on March 5, this picture is centered closer to the equator at 8 degrees south and at longitude 317 degrees. The diffuse markings are probably surface deposits of sulfur compounds and other volcanic sublimates. The dark spot with an irregular radiating pattern near the bottom of the picture is thought to be a volcano with lava flows.   (JPL)

small to be resolved on the images. While the eruptive plumes seen by Voyager were located in a 45-degree-wide equatorial band, where the shield-type volcanoes are also concentrated, several of the hot spots were observed in high latitudes. Abundant volcanic centers which are associated with obviously older deposits indicate that the polar regions were quite active volcanically not long ago. At latitude 70 degrees south, for example, there is a prominent volcano which is very similar to volcanoes on Earth and on Mars, and quite different from the volcanoes with erupting plumes. This volcano has a 30-mile-diameter central caldera which possesses several basins. There are dark lava flows radiating from the caldera. Close by is a smaller volcano of the same type.

The equatorial band of plumed volcanoes has deposited a wide equatorial band of lighter material obscuring the dark reddish material that is so evident closer to the poles. These dark regions are the dark polar caps identified in Earth-based observations.

sometimes or may be noneruptive because they lack sufficient volatiles to produce plumes. Most of the hot regions were dark. A few hot spots located in bright featureless regions may indicate that there are many small eruptive activities in those areas, too

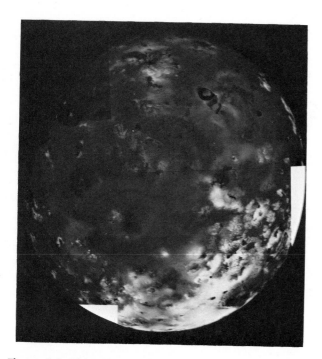

Figure 6-6: This picture, taken on March 4, is a composite. It shows two of the huge volcanoes; Loki at top right quadrant, and Pele at bottom left quadrant.   (JPL)

Within these polar caps are mountainous regions and layered regions, including some mesas revealed by erosion. Here there is evidence of long periods of deposition, faulting, and erosion.

Harold Masursky, US Geological Survey, at a colloquium on the satellites of Jupiter, stated that mountainous islands of suspected silicate materials are not extensive on Io, but they are widely distributed over the satellite. He commented that they may be remnants of vent rim deposits from earlier eruptions.

The surface of Io is undoubtedly young. No impact craters can be seen on it. Scientists have estimated that at the current rate of vulcanism the upper parts of Io's interior would have been recycled many times during the history of the satellite. It is the most geologically active surface so far discovered on any planet in our Solar System.

Nine volcanic eruptions were observed over the 6.5 days that Voyager 1 obtained pictures of Io. At least seven of the volcanoes were still active during the Voyager 2 encounter four months later. The volcanoes were originally given plume numbers 1, 2, 3, and so on, but were later given names: P1 is Pele, P2 is Loki (figure 6-6), P3 Prometheus (figure 6-7), P4 Volund, P5 Amirani, P6 Maui (figure 6-8), P7 Marduk, and P8 is Masubi. Between the two encounters a major eruption occurred and deposited a huge blanket of ejecta at 340 degrees longitude and 45 degrees north latitude.

The plumes appear to originate from long linear fissures as well as from concentric fissures surrounding calderas, and not from simple pipe vents like those of terrestrial volcanoes. Many other areas of Io exhibit surface features similar to those seen at the active sites (figure 6-9), and it seems reasonable to assume that these also are sites of eruptions which were not active at the time of the Voyager flybys but had been active recently.

There were major changes in surface markings between the flyby of Voyager 1 and that of Voyager 2. This implies that changes in color and albedo pattern occur over thousands of square miles of Io's surface within short periods. The magnitude of these changes implies that Io must be rich in volatiles such as the sulfur dioxide which is believed to be a major component of the volcanic activity.

If the interior of Io is molten to a large extent, it could be heated tidally to produce sufficient heat flow to drive the volcanoes. Existing ground-based measurements of infrared thermal emissions from Io have now been reexamined and hot spots have been discovered on them. The heat flow deduced from these observations is very large compared with the heat flows of Earth and Moon, and implies that the interior of Io must be partially molten on a global scale.

It is not clear whether sulfur or sulfur dioxide is the driving gas of the plumes, accelerating them to about 2,000 mph. This velocity could be obtained with sul-

a

b

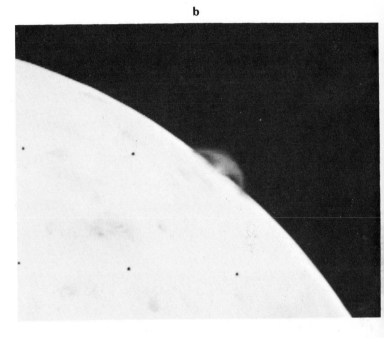

Figure 6-7: This sequence of two pictures shows the volcanic plume of Prometheus.

(a) Voyager picture taken about 12 hours before the spacecraft made its closest approach to Jupiter. The volcanic plume is the domelike structure at top right. (JPL)

(b) Taken just less than 2 hours after the first picture, this shows the plume rising from the limb of Io, displaying its fountainlike shape. The explosive force hurtling this material more than 60 miles into space is much more violent than any volcanic eruption on Earth, including Krakatoa. (JPL)

fur vapor when the sulfur is heated to more than 1000 degrees by molten silicates intruding into the sulfur. If sulfur dioxide heated by molten sulfur is the driving gas, there are problems of how thermal equilibrium might be maintained through phase changes and how frictional losses during acceleration of the gas are overcome. Some sulfur dioxide would certainly be expected to be present in all the plumes even if it were not the prime driver.

The material particles comprising the volcanic plumes of Io follow ballistic trajectories. However, the bright top of the plume from Pele may be evidence of a shock front stopping the particles and causing them to slide down to the surface along the shock instead of completing their ballistic trajectories to the surface. This may be because the particles ejected from Pele were traveling twice as fast as those in the other plumes.

The entire magnetosphere of Jupiter seems to be threaded by the orbits of uncharged sodium atoms speeding around the giant planet. Solar radiation affects the cloud of sodium ions to produce periodic east-west asymmetries in the spatial distribution and the intensity of the cloud. The Io plasma torus (see figure 6-10) acts as a good sink for sodium atoms near to the orbit of Io. It also produces short-term variations in the intensity of the cloud. From this it is estimated that there must be a strong flow of atoms from the satellite to provide those that must be needed to maintain the sodium cloud beyond the plasma torus.

Grand-based observations indicate that there are more sodium atoms in the cloud leading Io along its orbit than following the satellite. Ground-based images of the sodium cloud also show transient linear features. Both spacecraft were occulted by the torus during their flybys, so the signals from the spacecraft passed through the torus and provided the experimenters with much information about its characteristics. A study of the morphology of the hot plasma torus has shown that the brightness of the prominent line of ionized sulfur at 685 angstroms varies along the orbit. On the average these variations are cyclical, but they exhibit marked short-term differences from the average.

There is a sharp drop in the temperature of the plasma inside 5.7 radii of Jupiter from the planet.

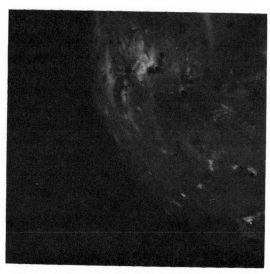

Figure 6-9: Another view of the lava flows and volcanic surface of Io. The smallest features are about 6 miles across. Many of the black spots on this and other pictures are thought to be volcanic craters; their temperature is higher than that of surrounding terrain. The complete lack of impact craters suggests that the surface of Io is much younger than that of the other satellites.   (JPL)

This is where the inner cold region is separated from the outer hot region of the torus. In the inner region ions are largely confined to the centrifugal equator, whereas the hot region is centered around the orbit of Io. The thickness is about 0.5 Jupiter radii. The ultraviolet spectrum of the Jovian system shows sulfur III and IV, and oxygen II and III ions. The emissions originate from the optically thin plasma torus centered at Io's orbit and aligned with the magnetic equator of Jupiter. The abundances of sulfur III and sulfur IV are about equal, and there are approximately equal numbers of sulfur and oxygen ions in the central region of this hot torus. In that region the electron temperature appears to be about 70,000 K, but there are also some electrons at about 1,000,000 K.

The northern and southern fringes of Io's sodium cloud are also observed to fluctuate in intensity. They are not in phase and seem to result from impact ionization from the plasma torus. This torus rotates relative to the sodium cloud because it is fixed

Figure 6-8: Two more Io volcanoes are seen shooting their plumes off the limb of a crescent Io. The top plume is the volcano Amirani, the one below it is Maui. This picture is one of about 200 images obtained during a 10-hour volcano watch. These images were placed in sequence to form a time-lapse motion picture of the volcanic activity on this satellite.   (JPL)

Figure 6-10: The Io double torus is depicted in this artwork showing the satellite Io emitting sulfur along its orbit and the generation of plasma regions because of interactions with the magnetosphere of Jupiter, as described in the text.   (JPL)

within Jupiter's co-rotating magnetosphere. The sodium cloud is spread out along Io's orbit. The plasma torus is also tilted about 8 degrees to the plane of the satellite's orbit so that only a part of the neutral torus is exposed to the intense ionizing radiation of the plasma torus. The fringes of the sodium cloud are ionized in the 1.5 hours that it takes the plasma torus to pass through the sodium cloud.

The Jovian decametric radio waves originate from Io's interaction with the magnetic field of Jupiter and with the plasma disc. Additionally, there is an effect from the interaction of the torus plasma with Io. Radio astronomy measurements made by Voyager revealed that the decametric radiation contains many individual features which have been called decametric arcs. These are believed to be produced by interaction of Io with the plasma torus. One explanation is that strong interactions occur when Io is in low-density regions near the northern and southern boundaries of the torus, and the decametric radiation is generated by multiple reflections from a standing Alfven wave current system which is excited by Io's interaction with the Jovian magnetosphere.

Near to the orbit of Io the atoms of the torus are uncharged, but they soon react with the plasma and become charged ions. Then they are affected by the magnetic field of Jupiter, spinning along with the planet and whirled into a co-rotating dance around it.

A significant difference was noted between the Io torus as measured by Pioneer and that measured by Voyager, which implies that the torus changes dramatically with time. One possible cause might be change in the volcanic activity of Io which supplies particles to feed the torus. The torus is also important in that its particles are believed to travel along magnetic field lines and produce the enormous aurora seen in Jupiter's polar regions.

Particles can be removed from Io by interaction of dust with the magnetosphere of Jupiter. Small grains of dust are ejected by the volcanoes to high altitude where they become charged by the Jovian plasma and accelerated free of Io's gravity. Such dust might fall toward Jupiter and collide with the ring particles, eroding large particles and breaking apart smaller particles.

The volcanoes of Io are not likely to be sending material directly into the magnetosphere of Jupiter because velocities in the plumes do not reach escape velocity from Io. However, the particles in the plumes could become electrically charged by the Jovian plasma, which would then permit them to escape from Io.

Although Jupiter would be expected to focus meteoroids and thereby cause Io to be more heavily cratered than the other Galilean satellites, no impact craters can be seen on the Voyager images of its surface.

Spectrophotometry from Voyager indicates that the surface of Io is more porous than that of Ganymede. The brightness of Io when observed at a Jovian opposition varies from year to year. It was bright during the years from 1973 to 1976 and fainter in the years 1977 through 1979. The changes in albedo which cause these changes in brightness are now believed to originate from the deposits laid down by the major eruptions.

As mentioned earlier, there are two schools of thought about the volcanic eruptions. One contends that these eruptions are driven by sulfur dioxide, the other that they are silicate magmatic eruptions.

Volcanic centers on Io are black spots, often surrounded by very dark halos as though from enormous deposits of ash. There are also many flows spread across the surface of Io and originating at the dark volcanic centers. Some are fan-shaped, others snake across the bizarre landscape for hundreds of miles. These flows favor colors of orange and red.

The great volcanoes are surrounded by markings that differ from the surrounding countryside in brightness and color, and change with time. At the time of the Voyager 1 flyby, the marking around Prometheus was a white doughnut-shaped marking. The one around Pele was shaped like a dark hoofprint. They are formed by the plume-shaped eruptions spreading material around the volcanic vents and fissures.

On Earth and the other terrestrial planets the heat source which keeps their interiors hot is believed to be the decay of the radioactive elements thorium and uranium. While undoubtedly there has been radioactive heating of Io, the major heating is believed to have resulted from tidal interactions between the satellite and the Jovian system. Io is only 262,230 miles from Jupiter. (Our Moon is on the average 238,800 miles from Earth.) Enormous tides are raised by Jupiter on Io, pulling it out of shape. However this tidal distortion, once fixed, would not result in continued heating of Io. But the other Galilean satellites play an important part in stimulating Io to volcanic activity. They also pull at Io, trying to snatch it away from or push it toward Jupiter. These tugs and pushes acting against the forces of Jupiter agitate the interior of Io and generate enormous quantities of heat.

The Jovian magnetic field at Io is very strong and its direction and intensity varies periodically. The variation induces an electric current in the interior of Io, which in turn heats the interior. Other heating occurs from current flowing into the satellite from the Io flux tube.

If this heating has continued since soon after the formation of Io it has had the important effect of differentiating Io gravitationally in a planetary refining process. Lighter materials have been able to float like dross to the surface, while denser materials such as metals have sunk to form a central core. Large abundances of sulfur on Io imply that the satellite must have initially possessed considerable quantities of water and carbon. All the volatile elements have escaped from Io into space. Water and carbon dioxide left Io millions of years ago. The satellite is dehydrated. Sulfur has floated to the surface and forms sulfur oceans beneath a fragile sulfur crust. The ocean may be several miles deep. Remaining oxygen has combined with sulfur to form sulfur dioxide, which exists as a liquid in a mud of solid sulfur and liquid sulfur dioxide floating below the crust on top of the sulfur ocean.

As radioactive and tidal heating rapidly depleted all Io's reserves of carbon dioxide and water, loss of hydrogen would be accompanied by oxidation, and continued convection would entrap oxidized materials deep within the satellite. As Io heated further, reactions occurred between the sulfur and the minerals rich in oxygen, thereby producing sulfur dioxide. Liquid sulfur was also formed by the heating process. Basalt rocks deep within the satellite must have been molten for several billion years, and vigorous convection in them continues until the present day. The escaping sulfur dioxide produces fissures in the sulfur crust through which liquid sulfur continues to erupt.

When the liquid sulfur dioxide expands under the influence of heat from the interior, it pushes up the plumes of sulfur and sulphur dioxide gas through vents and fissures in the sulfur crust. The sulfur and sulfur dioxide then fall back to the surface as snow.

Sulfur dioxide appears to be a major component of the white regions of Io and it is present in all other regions. The Voyager infrared (IRIS) experiment detected gaseous sulfur dioxide in the volcanic plume over Loki. It has been estimated that each plume ejects about 100 billion tons of material per year. Spectra obtained with the Ultraviolet Explorer Satellite show that sulfur dioxide frost is concentrated on Io at longitudes between 72 and 137 degrees, and is least abundant between 250 and 323 degrees. Sulfur dioxide seems concentrated in the white areas, and there is some other sulfur-based material concentrated in the brownish-red areas.

Io probably has a tenuous atmosphere of sulfur dioxide because Voyager measured gaseous sulfur dioxide above hot regions of Io. Pioneer 10 discovered an ionosphere, implying there must be an atmosphere, and the volcanic eruptions and presence of sulfur dioxide frost imply that gases must be present. However, Voyager 2 obtained two images of the same hemisphere of Io; one with the hemisphere illuminated by the Sun and the other in shadow but illuminated by reflection of sunlight from Jupiter. When the two images were compared there were no great differences in albedo, as might have been expected if sulfur dioxide had frozen out of the atmosphere at night when the temperature was 40 degrees lower than during daytime. It is concluded, therefore, that the atmospheric pressure on Io must be less than one millionth that of Earth's sea level atmospheric pressure. The large scale height of the ionosphere, coupled with the large molecular weight of the sulfur dioxide making up the atmosphere, suggest that the ionosphere is at a high temperature, which could result from the flow of current through the ionosphere induced by the co-rotating plasma of the inner Jovian magnetosphere.

There is an intriguing dynamic lock of the three bodies, Io, Europa, and Ganymede. The period of Io is half that of Europa, and the period of Europa half that of Ganymede. During evolution of the system it is probable that Io was driven out from Jupiter by the tidal forces until it came into resonance with Europa and later with Ganymede. The three satellites then slowly moved out from Jupiter, driven by the tidal forces on Io but locked together as a group by their own resonances.

Orbital motion of the Galilean satellites exerts control over their physical properties through tidal heating, and tidal interaction between the satellites and Jupiter has governed the evolution of the satellites' orbits. One theory is that the system was formed with the satellites out of resonance and they subsequently became locked into resonance. If this were so the tidal melting of Io would have been a comparatively recent occurrence. In this scenario Europa would not have melted. An alternative is that when the system formed the resonances were stronger than they are today. In such an evolutionary scenario Io would have rapidly melted very early during its lifetime and remained molten to today. Europa would also have a mantle of liquid water today, and Ganymede would have melted early and then frozen to the state it is in today.

The Galilean satellites have become very important in testing theories of the evolution of planets. There are problems in evolving theories for the origin of the Jovian system which fall into two main alternatives; Jupiter either formed much like a star does from instability in a massive primordial nebula, or grew upon a solid core by an accretion process collecting gases and planetesimals from the solar nebula. A system of satellites can be accounted for by either method of formation. Neither alternative, however, appears to lead to a system of satellites that would be locked in resonance at or quickly after formation of the system.

Europa

Europa is as different from Io as from Ganymede and Callisto. Early pictures of this satellite from Voyager 1 showed it as an almost featureless object, a bland whitish disk. The best pictures showed long linear features criss-crossing the satellite. From the Earth, Europa is a bright object, and observations had indicated that it must be an ice-covered world. And yet this satellite could not consist completely of water, for its density is relatively high among the Jovian satellites, second only to Io.

Voyager 2 was programmed to obtain close images of Europa, and spectacular images were obtained (figure 6-11). As the spacecraft moved toward the satellite the cameras first showed a featureless surface and then a planetwide system of cracks. Like Io, Europa is characterized by the absence of impact craters; only a few have been seen, and these seem to be fairly young. This implies that the surface of Europa, as with Io, is of relatively recent formation. Alternatively, the surface may be old but of a type on which impact craters may not survive for long.

Europa's unusually smooth surface is covered entirely with a confusing pattern of streaks, most dark, but some light. The streaks are very narrow, not more than 50 miles in width, and most of them are much narrower than this. Some of the streaks are curved, but most are straight. Although these lineaments look dark relative to the bright surface of Europa, they are actually only slightly darker than the surface. On the rest of the featureless white terrain there are some dark spots. Elsewhere there is a mottled surface. The streaks, dark spots, and mottlings are the major features discernible on this strange world (figure 6-12). There is some hummocky terrain associated with brown spots. One speculation is that the hummocky terrain represents icy material somehow anchored to a deep ocean floor. The low ridges are more numerous on the flat terrain than on the hummocky terrain.

A dark area which has been named Tyre Macula is about 90 miles across and is surrounded by concentric ridges. It may be evidence of an impact crater that is gradually losing its form in the fields of ice.

A strange thing about the streaks is that, though they look like cracks, they are not. In fact, they are depressions; they are more like the fracture pattern on an

a

b

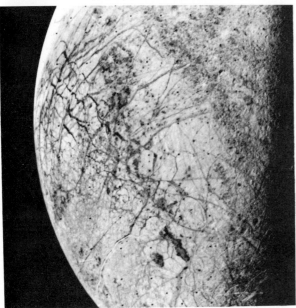

c

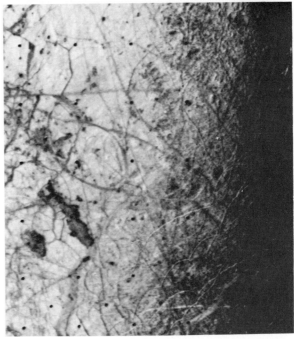

d

Figure 6-11: The strange world of Europa, dense but ice-covered.

(a) At 2.6 million miles the surface looks smooth with only a few indistinct markings but with linear features reminiscent of the drawings of canals on Mars by nineteenth-century astronomers. (JPL)

(b) At 750,000 miles the linear fracturelike features are quite plain. Europa begins to look more and more like a cracked egg. (JPL)

(c) This picture at 150,000 miles shows the cracks in great detail. Scientists thought these represent a fractured icy surface, perhaps 60 miles thick. (JPL)

(d) At 140,600 miles the smallest visible features are about 3 miles across. This image shows the evening terminator and a topography of complex narrow ridges in addition to the cracks. Also, there are diffuse dark bands running for thousands of miles. No strong evidence for impact cratering suggests that the surface of Europa, like that of Io, is a relatively young surface. The complex intersecting dark markings and bright ridges suggest that the surface was fractured and material from beneath welled up to fill the cracks. (JPL)

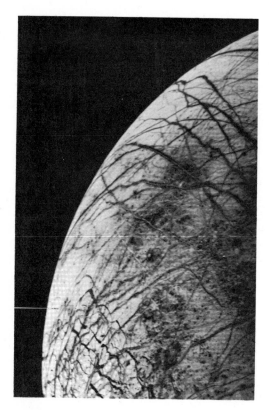

Figure 6-12: This high-resolution image of Europa's surface confirms its nature, a crust of ice about 60 miles thick overlying a silicate crust. The lack of relief, of any visible mountains or craters, is consistent with a thick ice crust; a veritable frozen ocean. The relative absence of features suggests that the crust is warm a few miles below the surface so that it easily deforms to its smooth shape. A tidal heating process might be responsible for keeping the interior of this satellite warm, as with Io, but heating the interior at a lower rate.   (JPL)

eggshell. The few light streaks (see figure 6-11d) are much narrower than the dark streaks, the widest being only about 6 miles wide. They also have some relief in the form of low ridges. These streaks are not straight either. They form curves.

The dark lineaments are thought to be fractures in the icy crust produced by a global distortion of Europa. There are two main types of these lineaments, a radial and concentric pattern located in a region centered at 20 degrees south and 170 degrees west, and sets of intersecting fractures elsewhere. The ra-

dial pattern may be caused by tidal deformation of Europa by Jupiter when one hemisphere had become locked to face Jupiter as it is today. This theory is supported by the fact that the center of this area of deformation is approximately radial to Jupiter. Whether the other fractures resulted from this same effect or from internal heating of Europa is still uncertain.

The consensus is that Europa has a planetwide ocean of water, a deep ocean that is covered with a relatively thin crust of ice. Heat rising from the interior of Europa cracks the ice crust, possibly moving it in plates.

Ganymede

The largest of the Galilean satellites, Ganymede (figure 6-13), is bigger than the planet Mercury, but it has a very low density, which implies that about half of the satellite consists of water. Speculation based

Figure 6-13: Ganymede as seen from 4 million miles away; quite different from Io and Europa.   (JPL)

on observation from Earth was that the interior of Ganymede consists of a silicate core extending some 1,000 miles from the center. Surrounding this core is a deep water ocean surmounted by a thin ice crust. The basic assumption is that the silicate core would develop sufficient radioactive decay heating to keep the water liquid, so that only the outer crust of perhaps tens to hundreds of miles' thickness would be ice. Moreover, the presence of vast quantities of liquid would enable the silicates to settle and reinforce the core. This theory would apply equally well to Callisto, the other low-density satellite of Jupiter.

A fascinating possibility would be for thicker parts of the crust to be supported like iceberg continents, much in the way that Earth's continents are supported. Moreover, movements of these iceberg continents might be analogous to the plate tectonics of Earth but driven by convection in the globe-encircling deep oceans of Ganymede and Callisto. Another conclusion was that impact craters would fairly rapidly deform and disappear.

The surprise from Voyager was that Europa, which is much denser than Ganymede and Callisto, appears to have the smooth icy surface from which nearly all evidence of impact cratering has vanished, while the two water-rich satellites show much evidence of impact cratering.

The surface of Ganymede is extremely diverse. Voyager provided images with exceptional detail showing the great dark area of Regio Galileo thousands of miles in extent (figure 6-14), inordinately complex systems of mountains and valleys, impact craters, lightly cratered smooth regions, jumbled mountains, ray craters, lateral faults, multiringed basins, and ghost craters.

Ganymede's surface is divided about equally between ancient cratered terrain and the younger grooved terrain (figure 6-15). Cratered terrain seems slightly more elevated than the grooved terrain. In the cratered terrain there are few very large craters or impact basins. It is believed that the ancient crust on which the impact craters were heavily implanted

was relatively thin, and this permitted the large craters to disappear through crustal collapse. Later the crust thickened and the grooved terrain of ridges and valleys was formed. These grooves appear to have been formed by faulting, and large polygons of the heavily cratered surface were rotated as the grooved terrain formed. The patterns are interpreted as resulting from an expansion of the crust producing uniform tensional stresses. This is borne out by the lack of regular patterns to the grooved terrain. The groove pattern may have originated from dyke extrusions between blocks of ancient cratered terrain, or grabens of cratered terrain which were later filled by extrusive floods. The fact that grooved terrain truncates older features suggests that the latter explanation is the correct one.

The grooved terrain consists of grooves which are spaced about 2 to 5 miles apart and have depths of about 1,000 to 1,200 feet and very gentle slopes. The grooves occur in sets that are sometimes fan-shaped, or bundled, or in overlapping patterns. Often they terminate at scarps or against other grooves. They cut across blankets of ejecta from craters and sometimes across crater rims, even when these craters are formed in the grooved terrain. This indicates that there must have been several episodes of formation of this type of surface.

The heavily cratered terrain implies that Ganymede must have experienced a bombardment from space early in its history like the inner planets of the Solar System. But it appears from analysis of the sizes and distribution of the craters on Ganymede that the population of impacting bodies differed from that in the inner Solar System.

The crust today appears surfaced with a regolith of fractured material on the heavily cratered and the grooved terrains. This regolith is, however, darker on the heavily cratered terrain and probably contains much debris from meteoroids.

There are young features on Ganymede, such as the rayed craters (figure 6-16). Bright rayed craters with dark floors may have been caused by the impact of

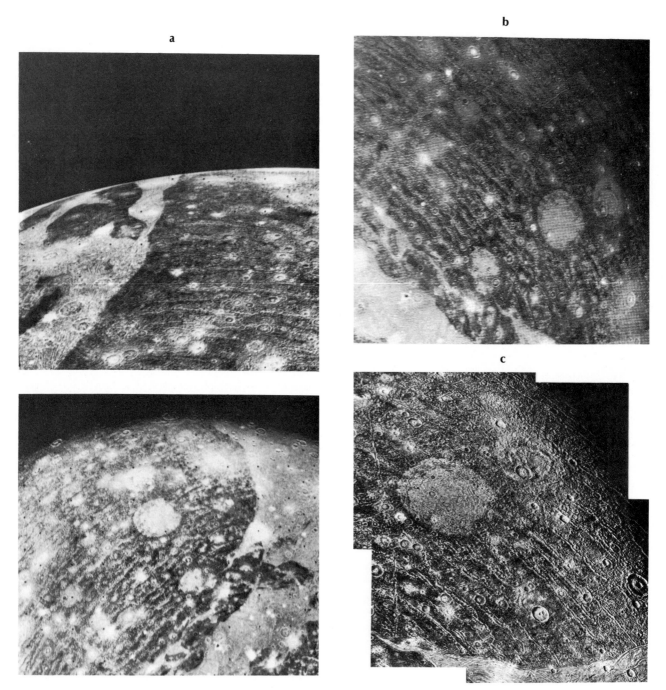

Figure 6-14: Ganymede has a large area of dark, heavily cratered ancient terrain which has been named Regio Galileo.

(a) The top image is centered at 60 degrees north latitude and 170 longitude, the bottom is at 20 degrees north and 130 longitude. The light, linear stripes recurring across the dark region are suggestive of some ancient impact basin of which there is now no other trace. (JPL)

(b) This heavily cratered terrain of Regio Galileo is believed to be the oldest part of the surface of Ganymede. The round features without relief are palimpsests, or ghost craters. All large craters are subdued in this way. (JPL)

(c) This close-up shows the big palimpsest in more detail, revealing its very low topography. (JPL)

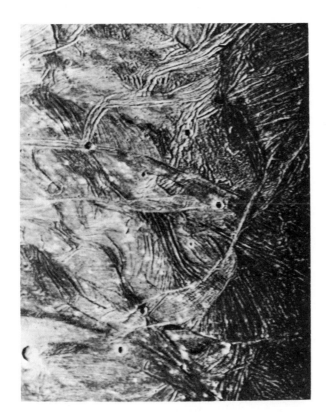

Figure 6-15: Grooved terrain on Ganymede imaged from a range of 87,000 miles. The width of this picture is about 350 miles and it shows features 1.7 miles across. Some of the systems of ridges and grooves are superposed, showing that there were different periods of formation of these features. A crater is cut by some of them, showing it was formed first. Other smaller craters were formed after the ridges and grooves.
(JPL)

cometary nuclei hitting the surface like giant snowballs. Ice spread wide over the surface to produce the bright ray systems which have not yet been darkened by a regolith. There are also dark rayed craters with bright floors which might have been caused by the impacts of small asteroids. However, there is a mystery in the fact that some craters have both bright and dark rays. Another mystery is that rayed craters are more common on Ganymede than on Callisto. The fact that there are more bright rayed craters in the grooved terrain of Ganymede suggests that this terrain has somehow become enriched with ice compared with the cratered terrain, since such a con-

centration of bright rays could not be explained on the basis of random impacts of cometary nuclei. Thus the rays might be icy material from the satellite itself.

Smooth terrains seem to have overlaid earlier features in thin swaths and narrow wedges. They are thought to result from active tectonics providing new surface material from below the crust. Thus the crust of Ganymede appears to have been reworked, at least in part.

On Ganymede there is a 160-mile-diameter dome that has been interpreted as evidence of water vulcanism. The summit of the dome rises about one-and-a-half miles above the surrounding plains and there is a field of many small craters surrounding it. One explanation is that the dome originated when impacts broke through a thin crust and allowed water to be released onto the surface, where it promptly froze until the vent was plugged. The dome would result if water came to the surface through a series of fractures in the crust rather than a single penetrating hole. A similar dome feature is observed at another location on Ganymede.

The ridges and grooves within the light albedo zones (figure 6-17) have been termed *sulci*: furrowed or grooved material. Close examination shows that these features were formed at several different times and not all in the same period (figure 6-18). There are also significant variations in crater counts within these terrains and within earlier features that have been modified by them. Ganymede's surface thus exhibits a complex history the causes of which we do not yet fully understand.

Craters of about 60 miles or more in diameter have an outer scarp, and an inner peak ring or a central pit, but craters with a central peak are not common except in the diameter range of 8 to 25 miles. The lack of central peaks in larger craters is puzzling when compared with the inner planets. A possible explanation is that the surface of Ganymede has changed by thickening of the crust. The pit craters are seen also on Mars, but not on the Moon and Mercury. It would seem that pit craters are charac-

b

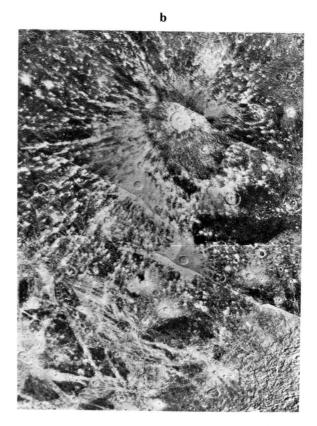

a

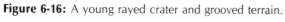

Figure 6-16: A young rayed crater and grooved terrain.

(a) At the bottom of this picture is a young rayed crater, Osiris. Above it the grooved terrain forms sulci. The dark material is ancient, heavily cratered terrain. (JPL)

(b) Osiris crater is at the top of this frame. There are other impact craters also with ray systems and believed to be relatively young. At the bottom of the picture portions of the grooved terrain transect others, indicating that they are younger. Osiris is about 93 miles in diameter. (JPL)

teristic of worlds with plenty of ice in the crust, as is believed to be the case for Mars.

Smaller craters are nearly all pit craters without high walls. The floors of the smaller craters are convex, indicating a slow viscous flow collapse after their formation. Because the central pits are larger in older craters of a given size, it is inferred that the crust of Ganymede must have changed in characteristics during the bombardment period.

While there are very few large craters, there are ghost craters which appear as circular features of different albedo (see figure 6-14c). These have been termed *palimpsests*; a term originating from parchments or tablets on which earlier messages have been erased and new messages written over them. A few of these have traces of crater walls.

Ganymede possesses only one large basin, Gilgamesh, which also suggests that the crust of the satellite has strengthened with time. Across the heavily cratered terrain there are regularly spaced grooves in wide arcs (figure 6-19). They are thought to represent rings surrounding a very ancient impact basin, similar to the concentric rings seen on Callisto. But there is no evidence of the actual basin on Ganymede today.

Ganymede has light polar caps, probably of frost. Photopolarimeter observations indicate there is snow at the north polar cap, and a mixture of snow and silicates in equatorial regions. But the water frosts on Ganymede appear to be mixed with a strong ultraviolet absorber. Infrared observations reveal a wide range of temperatures on this Galilean satellite. The poles at night are about −315° F, the cratered terrain at noon is about −190° F, and the grooved terrain is about −198° F. Such temperatures would permit the surface to consist of a deposit of silicate dust overlying a regolith of silicates and ice.

During the Voyager mission ultraviolet observations were made of the occultation of a star, k Centauri, by Ganymede. The data show that Ganymede cannot possess an atmosphere of oxygen or water vapor with a pressure greater than 2×10^{-8} millibars.

Figure 6-18: Located on the equator at 210 degrees longitude, Tiamat Sulcus consists of mutually intersecting bands of closely spaced parallel ridges and grooves. Near the bottom of the picture the grooves appear to be offset by another narrower band of grooves at right angles, as though by faulting.

Callisto

The outermost Galilean satellite, Callisto appears the least active. It is an ancient dead world on which, like our Moon, the surface contains the records of many ancient impacts from the early years of the Solar System. The surface has virtually no evidence of continued internal activity but is heavily cratered and has several large impact basins with concentric rings (figure 6-20). Callisto's density (1.8 times that of water) is similar to that of Ganymede, so Callisto is presumed to be a world consisting of about half water and half silicates.

The cratering on Callisto is quite dense with craters shoulder-to-shoulder in many areas, but there are few large craters. The silicate icy crust does not appear to be able to retain large craters because the ice flows slowly to fill crater floors. In striking contrast

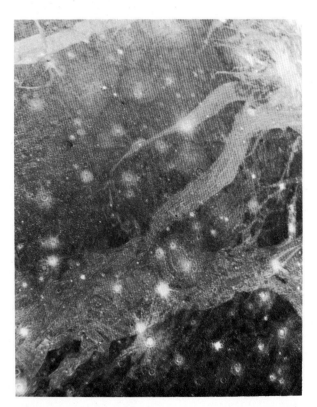

Figure 6-17: Sulci; grooved terrains, a peculiar feature of Ganymede. This picture is centered at 20 degrees north latitude and 200 longitude. It shows Mashu Sulcus, which has changed the feature of the heavily cratered terrain on which it lies. (JPL)

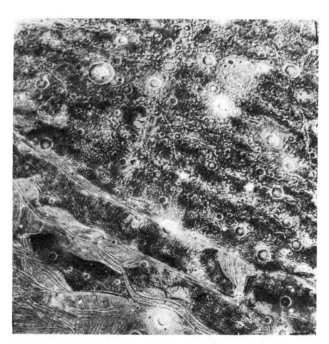

Figure 6-19: Sulci on the edge of Regio Galileo, across which are the wide arcs thought to be evidence of an ancient large impact. These features resemble circular ridges on Callisto that surround an almost completely annealed impact basin. The feature on Ganymede may be of similar origin, but all traces of the impact itself have vanished.   (JPL)

to Ganymede, very few rayed craters make bright stars on the surface of Callisto. The large number of older, nonrayed craters implies that Callisto has been geologically inactive for billions of years; quite different from the other Galilean satellites. The crust also does not support mountains. Even the big impact basins have been mellowed by the loss of vertical relief.

The biggest impact basin, Valhalla (figure 6-21), is located at 10 degrees north latitude and 55 degrees longitude. At its center is a 360-mile-diameter light-colored area with rays, surrounded by a series of concentric rings which may have been caused by differential loss of ice from water which welled up along fractures and froze. The rings are separated by dark material. Alternatively the ring system might represent shock waves frozen into the crust of the satellite. The ring system is asymmetric about the

central bright area. Valhalla does not have features like the lunar basins; there is no basin as such, no ring mountains, and no radial streamers of ejected material.

The surface of Callisto (figure 6-22) seems to be dark rock or regolith, but its exact nature is unknown. The leading hemisphere appears to have a surface that is similar in its microtexture to the lunar regolith. The trailing edge, however, has a coarser nature. But the Voyager pictures show both hemispheres covered equally with craters. Bright features may be ice or snow, but much evidence points to the surface in general being a silicate surface. During ages of geologic time ice could have evaporated from an ice-rock mixture, leaving behind the rocks to form a sur-

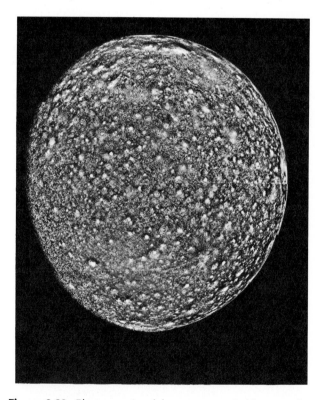

Figure 6-20: Photomosaic of the outermost Galilean satellite, Callisto. The distribution of impact craters is very uniform across the satellite. Near the limb is a structure, Ascard, resembling a giant impact basin. The limb, however, unlike that of Earth's Moon, is very smooth, confirming that no high topography is present on this world of ice.   (JPL)

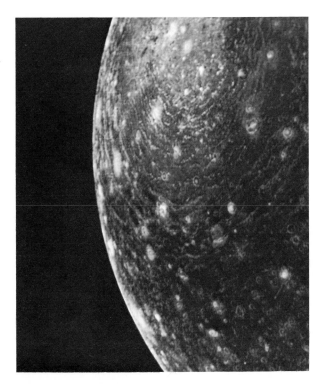

Figure 6-21: This picture of Callisto, taken by Voyager 1, revealed the great impact basin of Valhalla, about 400 miles across and with concentric rings extending to some 1,600 miles. It supports the concept that the surface of Callisto is extremely ancient and has not been molded by forces such as those at work on the other three Galilean satellites.   (JPL)

Figure 6-22: A computer-enhanced picture of Callisto shows that there are fewer young rayed craters than on Ganymede. Why is a mystery. The surface of Callisto is, however, the most heavily cratered of all the Galilean satellites and resembles heavily cratered terrains on the Moon, Mercury, and Mars.
(JPL)

face layer. Callisto is, in fact, the darkest of the Galilean satellites, but twice as reflective as our Moon.

Infrared observations tell us that the temperature on Callisto at noon is −180° F, and the lowest nighttime temperature is −315° F.

The Voyager mission had changed the Galilean satellites from indistinct globes with a few fuzzy markings on their surfaces into four distinct and bizarre worlds, each with a fascinating story to tell about our Solar Systems. This story will take many years to understand and its final chapters will not be written until Project Galileo can orbit Jupiter and study these strange satellite worlds over many months. In the more distant future we should hope to land automated spacecraft on their strange surfaces and analyze the chemical composition.

7.

The Rings of the Giant

IN addition to discovering the large satellites of Jupiter, Galileo made the first recorded observation of rings around a planet, although he did not interpret his observation correctly. Just before leaving Padua in 1610 to take up an appointment as professor at Pisa, Galileo directed his telescope at the planet Saturn. He recorded that the planet was in three parts; a large central disk accompanied on either side by two smaller disks. Later Galileo was even more mystified by his observations of Saturn when the two smaller bodies disappeared completely for a time.

It was not until 1655 that Christian Huygens correctly interpreted these observations and those of other astronomers who had been equally mystified by Saturn's changing appearances. Huygens had developed methods to grind lenses with much greater precision than his contemporaries, with the result that he was able to make much improved telescopes. He explained the peculiar appearances of Saturn as being

due to the presence of a very thin but wide ring surrounding the planet and inclined at a considerable angle to the plane of the ecliptic. From Earth the ring is sometimes seen open and at other times it is seen edge on. The open appearance resulted in Galileo's interpretation of three bodies forming the one planet. The closed situation was when Galileo was mystified by the disappearance of the two smaller bodies accompanying Saturn along its orbit. Huygens also discovered the large satellite of Saturn, Titan, one of the largest satellites in the Solar System.

For over three hundred years Saturn (figure 7-1) was regarded as a unique object in our Solar System, the only planet with rings. Then on March 10, 1977, two scientists were taking advantage of one of the rare occasions when Uranus occults a star. This occultation, which had been predicted in 1973 by G. Taylor, was of a faint star, SAO 158687, in the constellation of Libra. Such an occultation provides an

Figure 7-1: For many years Saturn was believed to be a unique object in the Solar System, the only planet with rings. This proved not to be true, and Saturn's rings themselves were shown by Pioneer and by the Voyagers to be different from what had been thought before these space missions. This picture from Voyager 1 in October 1980 shows Saturn and its ring system from a distance of 11 million miles. Three satellites are also shown. Outer left is Tethys; a little closer to Saturn on the left is Enceladus. Mimas is above and to the right of the rings. Unexpected radial "spokes" are seen in the rings at the left of the planet.   (JPL)

opportunity to measure the diameter of the occulting planet and the characteristics of its atmosphere as the light from the star is first dimmed by the atmosphere and then cut off by the planet. Robert Millis of Lowell Observatory traveled to Australia to observe the occultation with the International Planetary Patrol's 24-inch reflecting telescope located at Perth Observatory. James J. Elliot of Cornell University flew aboard the Kuiper Airborne Infrared Observatory, one of NASA's high-flying aircraft observatories that provided the advantage of being above much of the distortion-producing atmosphere of Earth.

Fortuitously there was some uncertainty about the time when the occultation would take place. So the astronomers started their observations early to make sure that the occultation would not be missed. Data was obtained about the intensity of light from the star for several minutes before the actual occultation. If they had not done so the rings might have been missed. It so happened that a true occultation did not take place at Perth; the star just missed going behind the planet. But when the results were inspected scientists were surprised to find strange diminutions of the light of the star before it should have gone behind the disk of the planet and again afterward.

The data from the high-flying aircraft experiment, from which a full occultation was observed, was even more revealing. It showed that the effects before occultation were a mirror image of those afterward. At first, when the Perth observers compared data with the high-flying observers soon after the occultation, it was thought that a swarm of small satellites may have been the cause, but further analysis led to the conclusion that Uranus, like Saturn, has a ring system (figure 7-2). Whereas Saturn's system was believed to consist of at least five, and possibly seven, broad thin rings separated by narrow gaps, the Uranian rings appear to be narrow and with wide gaps between them. The innermost ring is about 27,800 miles above the cloud tops of Uranus, the

outermost ring is 31,500 miles distant. Saturn, with its magnificent rings, was no longer unique.

Following this discovery, scientists began to look anew at the dynamics of ring systems and how rings might originate during formation of planets.

When Pioneer 11 flew past Jupiter some puzzling data were obtained about the radiation environment within the orbit of Amalthea. The spacecraft actually flew through a ring of Jupiter. Among theories put forward to explain the observations was that a small satellite, so far unobserved, might be circling Jupiter at about 1.83 Jovian radii. This could absorb particles and produce the unusual characteristics of the radiation pattern. It was also suggested that the absorber might not be a satellite but a ring of particles. Since this was before the discovery of the rings of Uranus, the idea of Jupiter having a ring was not very credible. Theories of celestial mechanics predicted that the tidal action of mighty Jupiter would quickly destroy such a ring or small satellite at that distance from the giant.

But Jupiter does have a ring. As recounted in an earlier chapter, Voyager 1 produced a discovery picture of this ring (figure 7-3) and Voyager 2 filled in more details (figure 7-4) and showed that the ring forward scatters sunlight very strongly. The ring is more easily seen from behind Jupiter than from the side on which the Sun is shining. The ring is extremely difficult to observe from Earth because it is thin and tenuous to the point of almost being transparent, so that it does not scatter sunlight back into the inner Solar System. Also, the brilliance of Jupiter masks the faint light from the ring.

A further clue in the puzzle of planetary rings would come when Pioneer and later Voyager flew by Saturn and showed that the Saturnian rings were by no means as simple as had been thought. Pioneer discovered an outer fine ring, the F ring (figure 7-5), and additional small satellites, and Voyager showed that all Saturn's rings are really complex systems of very many thin rings and gaps, some evenly spaced and others appearing randomly spaced (figure 7-6). Even the dark gaps in the Saturnian rings were seen to be

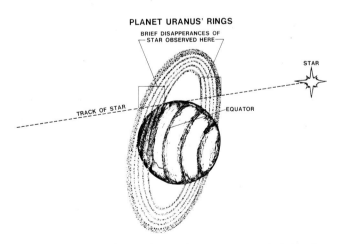

PLANET URANUS' RINGS

BRIEF DISAPPEARANCES OF STAR OBSERVED HERE

STAR

TRACK OF STAR

EQUATOR

Figure 7-2: Occultation of a star by Uranus revealed that that planet also has a ring system, but one that cannot be seen visually from Earth. It is hoped that Voyager 2, when it visits Uranus, will obtain images of this interesting ring system.

(NASA/Ames)

filled with faint rings. There is now much scientific speculation that all planets may have had rings during part of their evolutionary history. And Neptune may have rings today, yet to be discovered when the planet is explored by Voyager 2 toward the end of this century.

Voyager 2 was programmed to take more pictures of the ring during its flyby. These were taken when Jupiter was eclipsing the Sun so that they showed the ring brilliantly forward scattering sunlight when the glare of Jupiter was removed (figure 7-7). The ring particles forward scatter light extremely well. Unfortunately the highest-resolution pictures, which required also a relatively long exposure, suffered from smear because of the motion of the spacecraft during the exposure. So while there is a suggestion of structure within the rings, it cannot be identified clearly (figure 7-8).

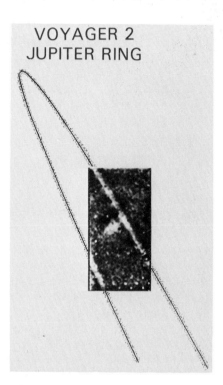

Figure 7-4: Voyager 2 looked again for the rings before and after its closest approach to Jupiter. This image was obtained July 8, 1979, at a range of 1.25 million miles. The spacecraft was 2.5 degrees above the plane of the rings. The picture shows that radially the ring is much narrower than individual rings of Saturn seen from Earth. Later we were to find that the rings of Saturn are made up of very many thin rings.   (JPL)

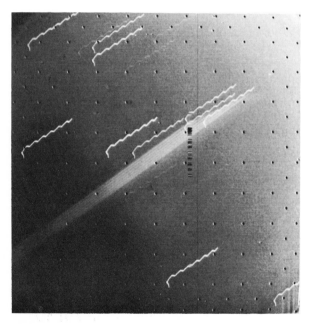

Figure 7-3: Rings around Jupiter were discovered on this image from Voyager 1 taken on March 4, 1979. The long exposure of 11 minutes and 12 seconds blurred the rings, which are extremely thin, and slight oscillations of the spacecraft during the exposure made the star trails wavy.   (JPL)

Examination of the data from Voyager indicates that the visible particles of Jupiter's rings are very small, their diameters measured in microns only; perhaps up to 5 microns. They are probably produced by collisions among larger bodies and even finer dust particles. Some of the dust particles probably originate from the volcanoes of Io.

The ring of Jupiter has three distinct components. There is the bright visible ring (figure 7-9), which is about 3,700 miles wide, with a single 500-mile-wide region of enhanced brightness assumed to be a concentration of particles in a ring within the ring. But no ring divisions are visible in the Voyager pictures. The ring is believed to be less than 1 mile thick. Its

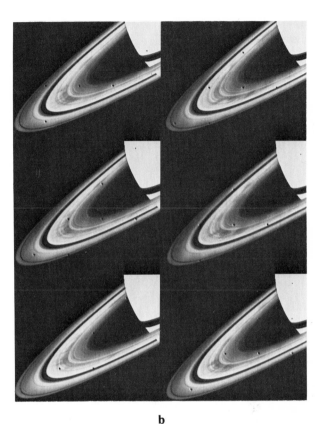

a   b

Figure 7-5: An inkling of what was to come: (a) Pioneer 11 obtained this image of Saturn's rings, which showed much structure within them and a very thin outer ring which was named the F ring. The large satellite on this picture is Tethys. The small satellite is one discovered by Pioneer.

(NASA/Ames)

(b) Dark spokelike features in Saturn's rings were revealed by Voyager and are shown in this sequence of images taken 15 minutes apart. (NASA/JPL)

outermost boundary is at 1.8 radii of Jupiter, or about 33,000 miles above the cloud tops. The total mass in the rings is estimated to be insufficient to make a satellite the size of Amalthea.

Some of the visible particles in the rings appear to spiral in toward Jupiter, forced out of the rings by radiation pressure of light from the Sun. From the inner edge of the bright ring there is a region which is much fainter than the main ring but extends down to the planet. This is believed to be a sheet of ring particles moving down to Jupiter to be destroyed as meteorites when each particle enters the atmosphere at high speed.

Surrounding the main ring there is a halo of material. It is extremely tenuous but extends at least 6,000 miles above and below the main ring. The Jovian plasma probably charges dust in the ring and the influence of the magnetic field of Jupiter will be to change the orbit of these dust particles so that they produce the observed halo surrounding the rings.

One theory for the continuance of a ring of Jupiter is that there are larger bodies within the ring which replenish the material, replacing that which spirals down to Jupiter. Although larger than typical ring particles, these bodies are too small to be seen in the Voyager images as individual satellites. Collisions

Figure 7-6: Voyager later revealed the great complexity of Saturn's rings system. This composite shows the many ringlets making up all the rings of Saturn, and the presence of ringlets in the gaps which from Earth seem dark.   (JPL)

Figure 7-8: These high-resolution pictures of Jupiter's ring were taken on July 10, 1979. The spacecraft was 2 degrees below the plane of the rings and about 961,000 miles from Jupiter. The forward scattering of sunlight reveals a radial distribution and density gradient of very small particles extending inward from the ring toward Jupiter. There is some hint of structure within the ring, but the motion of the spacecraft during these long exposures blurred the highest detail, particularly in the frame at the right.   (JPL)

among such bodies would lead to production of the fine particles making up the visible ring.

The external boundary of the bright ring corresponds with the orbit of the small satellite Adrastea (J14) discovered in the images from Voyager 2. This satellite probably defines the outermost edge of the ring system by preventing ring particles from passing outward beyond its orbit. Particles from Adrastea may contribute to the ring material. There is speculation that another small satellite may define the inner edge of the ring. But this satellite has not been detected.

Jupiter's ring, like the rings of Saturn, absorbs trapped radiation, as observed by Pioneer 11. Particles that intersect the ring will also be removed and some particles may be injected into the plasmasphere. These two effects lead to asymmetries in the inner plasmasphere of the giant planet.

How did the ring form? It has been proposed that the regular satellites of Jupiter condensed from a system of concentric rings created when the planet formed and shed excess angular momentum by ejecting rings as it contracted.

Figure 7-7: A brilliant halo around Jupiter, the thin ring of particles discovered by Voyager 1, is here portrayed by Voyager 2 from the night side of the planet. The spacecraft was deep in Jupiter's shadow so that the ring appeared unusually bright as it forward scattered sunlight toward Voyager. The planet is outlined by sunlight scattered in its atmosphere.   (JPL)

Figure 7-9: The Jovian ring is here drawn on a Voyager picture of Jupiter to show its extent and approximate dimensions.

(JPL)

The discovery of the rings of Jupiter, and the later discoveries by Voyager about the intricacies of the rings of Saturn, has raised many new questions in celestial dynamics and the evolution of planets. Rings associated with planets now seem to be more common, and there has even been speculation that Earth may have possessed a ring at one time during its evolution. It will be many years, perhaps decades, before these questions of how rings originate and how they fit into the general pattern of planetary formation and evolution are answered. It is hoped that, when Voyager encounters Uranus, and maybe Neptune, the spacecraft will provide new information to help us along the road to answering such questions.

8.

And Now the Survey

TO look, to be intrigued, to go. The journeys to the Jovian system and beyond have answered many fundamental questions about the giant planets and the outer regions of our Solar System. But as the earlier questions were answered, new questions of greater complexity were raised. This is the continuing story of the broadening of human consciousness. At first we believe we have simple explanations for what we observe. And then as observations improve we find that the simple explanations are marred by many flaws. We have to delve deeper and deeper as we seek a broadened knowledge of the processes and materials of nature. Our own place in a complex universe begins to clarify as we search for a crystal-clear jewel of understanding and perhaps the fundamental reasons behind a physical universe. The task so often seems to be never-ending. It very well might be, with the whole of creation spread before us waiting to be explored in space and time. The gauntlet is thrown down before us in many

fields; physics, chemistry, sociology, microbiology, genetics, as well as astronomy. Astronomers and planetologists are challenged to decipher the heiroglyphs of planetary surfaces which may explain in an as yet unknown language the story of their formation and the history of evolution. But the opportunity is ever present. It releases us from the humdrum of mere material existence, of providing food and shelter for ourselves as a higher animal species of planet Earth. It offers a new awareness of our ability to perform fantastic mental gymnastics as we project our minds without limitation, often through the sensors of complex machines and aided by highly sophisticated digital computers, into a galaxy of new environments and situations that we can never experience on our own planet.

Project Galileo (figure 8-1) is the next step in our exploration of Jupiter following the reconnaissance completed by Pioneers 10 and 11 and Voyagers 1

Figure 8-1: The Galileo Jupiter Orbiter spacecraft takes a close look at one of the Galilean satellites in this artist's rendering of the next mission to Jupiter. (JPL)

and 2. These spacecraft expanded our awareness of the Jovian system from indistinctly seen astronomical bodies to unusually strange worlds, and then to fascinating places: Io's sulfur dioxide fountains gushing into space from a land of fire and brimstone; Europa's cracked ice oceans; Ganymede's *sulci* stretching their furrows and grooves to distant horizons; and crater-strewn vistas of icy regolith on Callisto. And the spacecraft went on to Saturn (figure 8-2) and revealed previously unexpected strangeness in the ring system and the satellites.

But Galileo is more than a mission to Jupiter. It is an expression of the fundamental human trait of not being satisfied with a cursory examination and a simplified explanation, of wanting to delve deeper, of searching for what may, however, be an elusive fundamental truth. Coming twelve generations after Galileo Galilei first saw Jupiter in the newly invented telescope and wondered anew about the universe as bounded by the thought of the ancient world, Project Galileo is the important follow-up mission to explore

the giant of our system and its retinue of bizarre satellite worlds. It will also establish some new and powerful technology suitable for use in exploring the other fascinating outer planets, probing into the atmospheres of Saturn, Uranus, and Neptune, and perhaps landing on the utterly alien surface of Titan.

Project Galileo will orbit a spacecraft around Jupiter and plummet an instrumented entry probe deep into the swirling and colorful Jovian atmosphere. The aim is to seek information on the origin and evolution of the Solar System. Jupiter's atmosphere may be a sample of the original material from which stars are formed, still unmodified by nuclear processes. The Galileo Orbiter and an attached Atmospheric Entry Probe will be launched by the space shuttle in the mid-1980s, making use of one of the best launch opportunities during the remainder of this century for sending a large amount of scientific equipment into the Jovian system.

The instrument-laden probe (figure 8-3) will sample Jupiter's atmosphere for the first time. The orbiting

Figure 8-2: Comparisons between the giant planets and their systems are important to our understanding of the Solar System. Pioneer and Voyager provided an insight into the Saturn system, but a Saturn orbiter probe is also required. Particularly a probe is needed to penetrate the strange atmosphere of Titan, the large satellite of Saturn. (JPL)

Figure 8-3: Probe technology has been proved with the Pioneer Venus probes. A more advanced probe design will be used with the Galileo project. In this artist's rendering, the red-hot nose cone separates from the probe as it hangs on a parachute and samples the atmosphere of the Solar System's largest planet. (JPL)

spacecraft will use a gravity assist from two of the Galilean satellites to trace a series of elliptical orbits gradually moving the closest approach—the periapse—of the spacecraft's orbit around the planet. These orbits take Galileo Orbiter far out into the magnetosphere and then whip it inward for close encounters with Ganymede, Callisto, and Jupiter.

Because Galileo will orbit Jupiter for many months, it will obtain information that cannot be obtained by flyby spacecraft such as Pioneer and Voyager. Scientists expect Galileo to answer many questions that can only be answered by orbiting Jupiter and sending a probe into its atmosphere. The great interest in the mission among scientists brought 500 applicants wanting to participate, including 90 from foreign countries. Only 100 could be chosen. Galileo will provide for our knowledge of Jupiter what Pioneer Venus Orbiter and Probes did for our knowledge of Venus, a quantum jump forward in understanding another planet.

Scientific investigations are widespread in their scope. For example, meteorologists are intrigued by the processes taking place in the Jovian atmosphere, and how they differ significantly from the atmosphere of Saturn (figure 8-4). Among the basic and economically vital challenges in terrestrial science is to understand long-term weather and climatic changes on our planet. Explanation of the giant, long-lasting storms of Jupiter and how and why they differ from storms on Saturn, coupled with our knowledge now of Martian and Venusian meteorology, are expected to contribute to a better understanding of Earth's climate and long-term weather systems to which our daily lives are so intimately bound.

The Galileo Orbiter will look very closely at the big satellites. Voyager showed that all the Galilean satellites are unusual worlds, even more unusual when compared with the satellites of Saturn (figure 8-5) and the inner planets of the Solar System. Galileo will look very closely at at least two of the big satellites, inspecting the surfaces, measuring their magnetic and gravitational fields, studying their surface features, and finding out how the satellites interact with the strange environment of the Jovian magnetosphere.

In general the aim of the Galileo mission is to find out more about the chemical composition and physical state of the atmosphere of Jupiter, to study the surfaces of the big satellites, their composition and physical state, and to map the magnetic field of Jupiter and determine how energetic particles behave within it. The probe will determine the pressure, temperature, and density of the Jovian atmosphere to a depth where the pressure is about ten times that at sea level on Earth. At that level scientists believe the probe will have penetrated below the Jovian clouds and will be looking into the many thousands of miles of clear hydrogen. In passing through the clouds the probe will determine the location and structure of the various layers and how the constituents of the atmosphere vary at different levels.

a

b

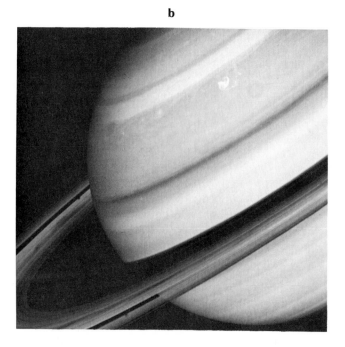

c

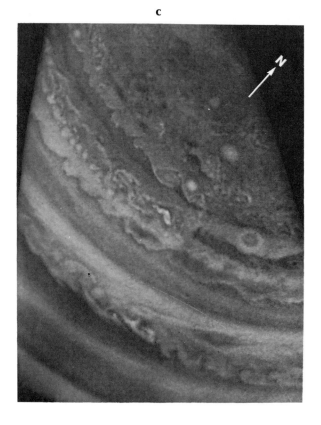

Figure 8-4: We can now compare the planets Jupiter and Saturn, and soon will be able to compare them with Uranus if all goes well with the Voyager 2 mission.

(a) and (b) Equatorial regions compared; the markings on Saturn are much more subtle than those on Jupiter. This is believed to result from a much deeper atmosphere above the clouds of Saturn.   (JPL)

(c) and (d) The polar regions of the two planets appear very similar; Jupiter on the left and Saturn on the right.

(NASA/Ames and JPL)

(e) and (f) There are belts and zones on both planets, those on Jupiter, top right, are more clearly defined than those on Saturn, bottom left.   (JPL)

(g) Circulation around spots in Saturn's atmosphere is also similar to that around Jupiter's spots, as shown in these two pictures by Voyager 2. The top picture was taken about 10¼ hours before the bottom. The motion is revealed as anticyclonic.   (JPL/NASA)

Project Galileo will be the first interplanetary mission to take advantage of the shuttle space transportation system. A shuttle will carry the Galileo spacecraft and a booster stage into orbit, from which the booster stage will propel the spacecraft on a path to

d

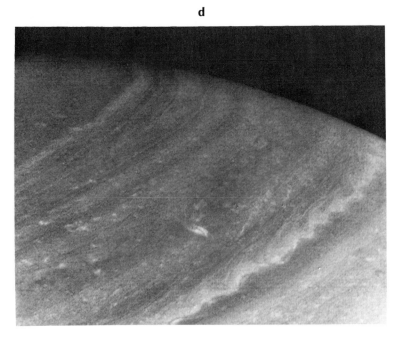

e

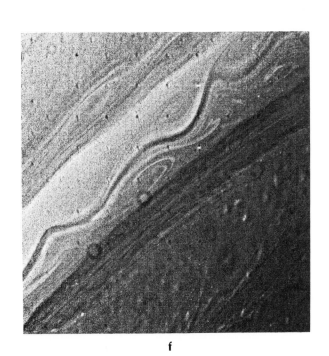

f

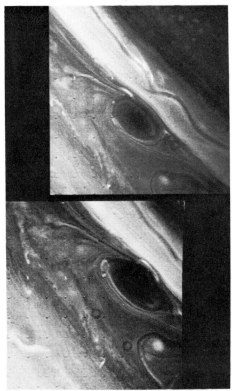

g

a

b

Figure 8-5: Saturn's satellites are also a special class of bodies intermediate in size between the big Galilean satellites and the small Jovian satellites. The satellites of Saturn are also more Moonlike with heavily cratered surfaces; the exception being Titan, which is shrouded from view with a dense atmosphere.

(a) Mimas; heavily cratered and with a huge impact on the hemisphere not seen on this picture. (JPL)

(b) Dione; again heavily cratered and with sinuous valleys and fault systems. (JPL)

(c) Rhea; with craters standing shoulder-to-shoulder like the heavily cratered lunar highlands. Rhea is the most heavily cratered of Saturn's moons. It also has multiple ridges and grooves similar to the Moon and Mercury. (JPL)

(d) Titan; its surface completely hidden by a thick atmosphere; a satellite of mystery. A dark polar cap was seen by Voyager, and there is an equatorial separation between the two hemispheres, one being darker than the other. (JPL)

(e) Tethys is very heavily cratered, but with part of the ancient surface partially reworked by internal activity. The large crater (upper right) almost lies on a huge trench system that girdles nearly three-quarters of this Saturnian world. (NASA/JPL)

(f) Enceladus presents the enigma of a surface that has been relatively recently remolded. Its morphology is closer to that of Jupiter's Ganymede than any other satellite. (NASA/JPL)

(g) The outermost of Saturn's large satellites, Iapetus, shows a heavily cratered northern hemisphere with the north pole close to the large crater. The leading hemisphere is mysteriously covered by some unknown dark material. (NASA/JPL)

(h) This composite shows seven of the very small satellites of Saturn. From top left to bottom right these are: Dione trojan, 1980S3, Tethys trailing trojan, 1980S1, Tethys leading trojan, 1980S26, and 1980S27. The last two are F-ring shepherding satellites. (NASA/JPL)

c

d

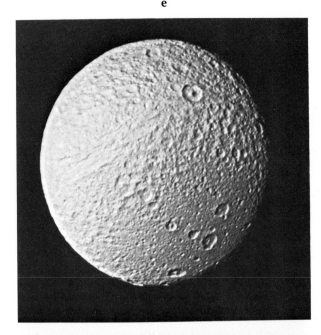

e

f

g

h

Jupiter. About 100 days before reaching Jupiter, the probe will separate from the orbiter and be aimed to enter the Jovian atmosphere slightly south of Jupiter's equator to sample the highest cloud regions. The Orbiter's path will then be changed so that the spacecraft will avoid hitting Jupiter prior to entering an orbit. Shortly before entering orbit it will observe the entry of the probe and relay data from the probe to Earth. The Orbiter's rocket engine will be used three times: first to put the spacecraft on a path from which it can observe the probe's entry, then to slow it into an orbit around Jupiter, and the third and final time to place it in an elongated elliptical orbit. From then on the perturbations of the Galilean satellites will be used to change the orbit, which will be timed so that the spacecraft periodically approaches Ganymede and Callisto. The orbital motion and gravity of these worlds will control subsequent orbits without further need of rocket propellant.

Space navigators will be able to use these pulls to steer the orbit around Jupiter to regions that would otherwise have been inaccessible. A close encounter with Io can be obtained, and the mission is extremely adaptable so that new discoveries can be thoroughly checked.

In a period of 20 months, the Orbiter will make as many as 12 orbits of Jupiter, and 11 encounters with Ganymede and Callisto. A final looping orbit will be directed into the magnetotail of Jupiter, where the magnetosphere stretches deep into space away from the Sun. After the nominal mission the spacecraft will continue into an extended mission if all its vital systems still function correctly. Most spacecraft launched in the last decade have continued into extended missions far beyond the original expectations.

The Orbiter incorporates many of the advanced technologies developed for previous interplanetary missions. It consists of two major parts, one of which can be despun while the other is spinning. The dual nature maximizes the effectiveness of two groups of experiments. To investigate the magnetosphere of Jupiter, scientists require a spinning spacecraft like

Pioneer to sweep the viewpoint of their instruments around in space. But imaging of Jupiter and the satellites requires a stable platform, like Voyager, from which instruments can be accurately pointed. The two parts can spin together during interplanetary cruise and when the spacecraft is maneuvering. A command from Earth despins the imaging part of the spacecraft when required for imaging experiments.

The spacecraft needs a large antenna for communicating the voluminous data that will be gathered by its many experiments. This antenna is folded for launch and afterwards extends like an umbrella to its 16-foot diameter. A new electrical power unit uses selenide converters to change heat from radioisotopes into electricity more efficiently than the earlier RTGs.

The Orbiter has a new camera system using solid state charged couple devices similar to those used on the most recent Soviet probes to Venus instead of the vidicon tubes flown on the Voyager mission. This allows pictures to be obtained over a wider range of the visible spectrum than was possible with Voyager. With the new system experimenters expect to be able to see details as small as 10 to 20 feet across on the surfaces of the satellites.

A near infrared mapping spectrometer will identify materials on the surfaces of the satellites, and a dust-measuring instrument will determine the size, speed, and electron charge of micrometeorites near Jupiter and its satellites. In addition, the Orbiter carries particle- and field-measuring experiments, and ultraviolet and radio experiments similar to Voyager.

The probe carries its instruments in a compartment behind a heat shield. To equalize pressure within the probe despite the pressure at different depths in the Jovian atmosphere, the probe is designed with vents so that the inside of the probe is always at the same pressure as the ambient atmosphere. This keeps structural weight down, so that the heat shield could be made more effective and more instruments could be carried than in a pressurized probe.

The Probe carries temperature- and pressure-sensing instruments. In addition a mass spectrometer determines the composition of the Jovian atmosphere. An interferometer determines the ratio of hydrogen to helium accurately to check on the measurements of the mass spectrometer. A nephelometer determines the sizes of particles in the atmosphere and the location of cloud layers, and a net-flux radiometer compares energy being received from the Sun with that being radiated from Jupiter at various levels in the atmosphere. Finally, a lightning detector will investigate the enormous lightning bolts that Voyager discovered in Jupiter's atmosphere.

By Jupiter might best be summarized in the words of Johannes Kepler, who had been so impressed by Galileo's discovery of the complex Jovian system of satellites, as recounted in *Conversation with the Starry Messenger*:

> There will certainly be no lack of human pioneers when we have mastered the art of flight. . . . Let us create vessels and sails adjusted to the heavenly ether, and there will be plenty of people unafraid of the empty wastes. In the meantime, we shall prepare, for the brave sky-travellers, maps of the celestial bodies— I shall do it for the Moon, you Galileo, for Jupiter.

But our Solar System is by no means the end of our quest for knowledge of the ultimate reality. Wrote astronaut Alfred M. Worden, who orbited that Moon which had been so painstakingly mapped from Earth by the ancient astronomers:

> The next step is out there. Out there stars shine, pieces of light . . . a pattern of so much brilliance that I am honored even here.

We are still undecided whether planets are common to all star systems. If they are, one of the great challenges of the future will be to seek out and explore planets beyond our Solar System and ascertain whether or not they have evolved similarly. Since we have not found conclusive evidence of living things elsewhere in our own Solar System it will be inevitable that in our quest for the meaning of life we shall seek it in other star systems. The Pioneers and Voyagers carry messages to the stars. Our next technical challenge will be to send unmanned probes to other star systems; probes able to send information back across the light years of interstellar distances. In anticipation of this period I once wrote:

> Man can see the glory of the stars, and his present technology if applied wisely can bring the exploration of other worlds within his grasp. . . . It would be rash to prophesy dogmatically upon these matters, but it would appear feasible that those same ever-accelerating developments in technology which are allowing Man to reach out into the Solar System will, if he grows to maturity without accidents, let him also achieve interstellar travel. (*Satellites and Spaceflight*, 1957)

As Man is poised with his machines breaking through the outer limits of the Solar System, those words are equally true today.

Epilog

NOVEMBER 16, 1971: Pioneer 10 gleamed under the brilliant lights of an artificial Sun in the harsh vacuum of the space simulator at TRW Systems, Redondo Beach, California. The spacecraft was in final test before being shipped to the Kennedy Space Center at the end of December. A group of science correspondents from the national press who were in Los Angeles for the arrival at Mars of the first orbiter spacecraft, Mariner 9, were at TRW Systems for a briefing on Pioneer. After being briefed by project management and principal investigators they had been invited to see the spacecraft under a test that was simulating conditions at six astronomical units from the Sun (figure E-1).

I was among those correspondents. When I looked at Pioneer through the portholes of the simulator, I suddenly became aware that this small but intricate machine would actually leave our Solar System. In several earlier books I had developed the idea of sending probes to search for planets of other star systems. Now I visualized how Pioneer 10 escaping from the Solar System would become mankind's first emissary to the stars. I thought of the importance of the spacecraft carrying a special message from mankind. Not the name of a president, or the names of politicians, or even the name of the fabricator of the spacecraft, all of which are quite meaningless in a cosmic context. Instead it should carry a message that would tell any finder of the spacecraft a million or even a billion years hence that planet Earth had evolved an intelligent species that could think beyond its own time and beyond its own Solar System.

The few people with me on the platform of the simulator did not see the importance of the idea when I mentioned it to them. But a little later at lunch I mentioned the idea to Richard Hoagland, a freelance

Figure E-1: Space simulator at TRW Systems, Redondo Beach, California in which the Pioneer spacecraft were tested before embarking on their endless journey into space.

(TRW Systems)

and Drake are the two people who are the most likely to be able to get the message on the spacecraft.

So with Hoagland I approached Carl, who was still visiting the Jet Propulsion Laboratory, Pasadena, in connection with Mariner 9's mission to Mars. Carl's eyes lit up as I explained the idea to him. With his usual enthusiasm he then talked with Frank Drake, who was also enthusiastic about the idea. The two of them designed a message, and Linda Salzman Sagan prepared artwork for a plaque to carry this message. The idea was then presented to the National Aeronautics and Space Administration. I heard nothing further about the idea until mid-December, when I was in Boston appearing on a panel with Bob Cowen of the *Christian Science Monitor* and Homer E. Newell, NASA Associate Administrator for space sciences. Homer told me that the plaque was on the spacecraft but there would be no publicity until the launch. Here I was with a story of mankind's first message to the stars and unable to publish! But it was wise to keep the plaque a secret, as some of the antiplaque stories that erupted later showed. Actually, when the story finally broke it proved to be one of the most widely used stories about a spacecraft, generating thousands of articles.

The plaque design was etched into a gold-anodized aluminum plate, 6 by 9 inches, and 0.05 inches thick. The plate was attached to the antenna support struts of the spacecraft in a position where it would be shielded from erosion by interstellar dust (figure E-2).

When Pioneer 10 flew by Jupiter it acquired sufficient kinetic energy to carry it completely out of the Solar System. About 40,000 years hence it will reach the distance of the nearest star, heading in the direction of the constellation of Taurus, the Bull. Somewhere between one and ten billion years from now it may pass through the planetary system of a remote stellar neighbor, one of whose planets may have evolved intelligent life. If that life possesses sufficient intelligence to detect the Pioneer spacecraft—needing a higher technology than mankind

writer, and to Don Bane, then with the *Los Angeles Herald-Examiner*, and they enthusiastically agreed on the need for such a message.

But how to get it on the spacecraft? Obviously if a few science writers approached NASA they would be unlikely to convince the right people that a message should be placed on the spacecraft. So who could do this?

A few days earlier I had dined with Carl Sagan at the home of Merton Davies, who was entertaining a mutual friend from England, Charles Cross. Carl had related his experiences when he and Frank Drake had attended an international conference in the Crimea to design a radio message suitable for transmitting to extraterrestrials. Obviously, I thought, Sagan

Figure E-2: A plaque with a message from humanity to extra-terrestrials, as first suggested by the author, was carried by each Pioneer spacecraft. It was mounted on the support struts to the big antenna with its message facing inward to protect it from erosion by cosmic dust particles. (NASA/Ames)

possesses today—it may also have the curiosity and the technical ability to pick up the spacecraft and take it into a laboratory to inspect it. Then the plaque with its message from Earth people should be found and perhaps deciphered.

A similar plaque was carried by Pioneer 11. After that spacecraft's encounter with Saturn, it too possessed sufficient velocity to escape from the Solar System but in almost the opposite direction from Pioneer 10.

The plaques on the two spacecraft are identical. Each tells of Man, where and when the species lived, and its biological form. At top left of the plaque (figure E-3) is a schematic of the hyperfine transition of neutral atomic hydrogen—a universal "yardstick"—that provides a basic unit of both time and physical length throughout the physical universe. As a further size

check, the binary equivalent of the decimal number 8 is shown between the two marks indicating the height of the two human figures. Then height can be compared with the scale of the spacecraft itself, which is also drawn on the plaque.

The hydrogen wavelength—about 8 inches—multiplied by the binary number representing 8 alongside the woman, gives her height, namely 64 inches.

The radial pattern to the left of center of the plaque represents the position of the Sun relative to 14 pulsars and the direction from the Sun toward the center of the Galaxy, which is indicated by the long horizontal line with no binary digits on it. The binary digits on the other lines denote time—the period of fluctuation of each pulsar. An extraterrestrial should be able to deduce this because the marks represent in binary form a precision equivalent to 10 decimal digits, which is unlikely for distances to stellar objects but quite feasible for measurements of time. From the unit of time established from the hydrogen

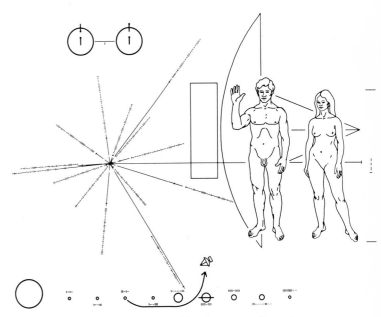

Figure E-3: The message carried by the interstellar stela was developed by Carl Sagan and Frank Drake. It is described in the text. (NASA)

atom, the extraterrestrial intelligence should be able to establish that the times represented by the binary digits are all close to one tenth of a second, which is typical for the period of a pulsar.

Since the periods of pulsars change at known rates, they act as galactic clocks. An advanced civilization should be able to search galactic records to identify the star system from which the Pioneer spacecraft originated, even several billions of years from now.

Below the orientation diagram another diagram identifies the relative distances of the planets in binary format and identifies the bright-ringed Saturn and the Earth from which the spacecraft originated. The path is shown past the largest planet and the spacecraft is depicted with its antenna pointing back to its origin on the third planet.

Finally the plaque shows line drawings of a man and a woman. The man is depicted in a characteristic gesture of human friendliness. This gesture shows how limbs can be moved and displays the important four fingers and their opposing thumb. The figures and physiognomy were carefully drawn to be ethnically neutral. No attempt was made to explain what may, to an alien intelligence, be mysterious differences between the two physical types—male and female.

The Voyager spacecraft will also escape the Solar System. These spacecraft carry identical messages to extraterrestrials. The Voyager messages are much more complex than the simple plaques of the Pioneers. Under the leadership of Carl Sagan, they were designed by a team of scientists and educators. The message is engraved on a 12-inch disc of copper together with electronic information that an advanced technological civilization might be expected to convert into diagrams, pictures, and printed words, including messages from the President of the United States and the Secretary of the United Nations, and greetings in about 60 human languages. A phonograph record was chosen because it can carry much more information than a simple plaque. Each record is protected by an aluminum jacket.

The record contains in scientific notation information on how the record is to be played, using a cartridge and needle provided. The record begins with 115 photograhs and diagrams in analog form, depicting mathematics, chemistry, geology, and biology of the Earth, photographs of people from many countries, and some hints of the richness of our civilization. There are schematics of the Solar System, its dimensions, and its location in the Galaxy, descriptions of DNA and human chromosones as we now understand them, photographs of Earth, the Voyager launch vehicle, a large radio telescope, and people in various activities. Next there are spoken greetings and a sound essay on how we believe the Earth and its life evolved. There are familiar terrestrial sounds of weather and surf, of animals and machines, of rockets and babies. The record also contains musical selections which are representative of the cultural diversity of Earth.

Whether or not an extraterrestrial will be able to understand these terrestrially oriented pictures and sounds that may require a terrestrial background for their interpretation is not important. The important point is that we have placed our civilization on record. Barring a collision course with a star or interstellar object—an extremely improbably event—these records of humankind will be preserved in space for billions of years.

As an epilog to our first missions to the giant planets of our Solar System, the plaques and the records are more than cold messages to an alien life form in the most distant future. They signify a strength of people of our planet, that in an era when troubles of war, pollution, clashing ideologies, and serious social and supply problems plague them, people can still think beyond themselves and have the vision to send messages through space and time to intelligent beings on star systems that have perhaps not yet condensed from a galactic nebula.

These messages are intellectual cave paintings, marks of Man, that should survive not only the red-giant destruction of all the caves of Earth when the Sun expands to engulf our planet, but also the ultimate end of the Solar System itself. They display mankind's spiritual insight and imagination, which it is hoped will overcome the material problems of this period of human emergence. And if not, if we descend into fossildom without traveling physically to the stars, we will at least have attempted to leave a record for other life forms of our Universe. Let us hope that no technologically more advanced civilization of our future will seek and destroy these emissaries in the name of some existing or new ideology.

Index